Reggie D. Millette

Hostas

Millette
photomedia

Book design by: Annie Duperré

Les Éditions du Sommet inc.

407-D, rue Principale
St-Sauveur-des-Monts (Québec)
J0R 1R4 CANADA
450 227-8668
info@dusommet.com
www.dusommet.com

Millette photomedia

62, 7e avenue
St-Ambroise (Québec)
J0K 1C0 CANADA
1 888-MILLETTE
info@millettephotomedia.com
www.millettephotomedia.com

ISBN: 978-2-89696-044-6

Printed in Canada by Friesens

Reggie D. Millette

Hostas

Grow your Hostas as per species habits

Millette
photomedia

Contents

PHOTOGRAPHIES

Most of the pictures (415) were taken by the author with a Canon 60D cameras.

Few photos were supplied by Ralph Heiden (14), Albert Mondor (2), Rock Giguère (1) Majella Larochelle (1) and Mark Zilis (1)

All the pictures were taken directly in gardens and are not touched up. So the pictures represent how they should grow in your garden.

The hosta picture on cover is *Hosta* 'Niagara Falls'

From the same author...

In French

Les Hostas, published in 2003 by Les Éditions de l'Homme

Les Hémérocalles, published in 2005 by Les Éditions de l'Homme

Les Iris, published in 2007 by Les Éditions de l'Homme

Les Hostas, published in 2013 by Les Éditions du Sommet

ACKNOWLEDGMENTS

The writing and realization of this book were possible with the help of many people.

Thanks to my partner, Claudette, who also loves hostas, which is also my great passion. She always encourages and supports me. She inspires me, she is my first critic and my first proofreader.

Thanks to Denyse Galarneau and Francine Labonté who verified all the texts.

Thanks to Mark Zilis who always helped me with all the details about these great plants. He also helped me compiling the hostas being part of the species and their growing habits.

Thanks to the Montreal botanical Garden, to Director Gilles Vincent, and to all members of the personnel for their tolerance and helps.

Thanks to the American Hosta Society who allowed me to access their register for the registered hostas.

Thanks to all the gardeners, for letting me take pictures in their gardens and their landscaping.

And finally, thanks to my parents and friends, and all the ones who helped and supported me.

In this landscaping, Hostas 'Blue Angel' and 'Just So' offer a nice contrast.

FOREWORD

The author Reggie D. Millette and Mark Zilis

Why is understanding a *Hosta* species so important to collectors and gardeners? First, and foremost, the adaptation of a species to specific environmental conditions in the wild has a direct correlation to how it will grow when brought into cultivation. The light, fertility, moisture, and temperature extremes found in a species native habitat offers a blueprint for its success in the garden. Though the majority of hosta species are adaptable to most shade gardening conditions, some are more tolerant than others. Understanding a species also allows for the recognition of the genetic traits that its offspring will display. Plant size, growth rate, and flowering habits are the most obvious, but disease resistance, durability, and breeding potential also fall along species lines.

Hosta plantaginea was the first cultivated species. It originated in southern China where the climate is fairly warm and humid. Not surprisingly, in cultivation *Hosta plantaginea* needs bright, warm conditions with plenty of moisture to thrive and produce its large, fragrant white flowers in late summer. Plant it in a too densely shaded area, especially in cooler climates, and it struggles to bloom. Adequate moisture is particularly critical for successful flowering in 'Aphrodite', a double-flowered form of *H. plantaginea*. Without sufficient rain during the growing season, its flowers do not open properly. Many of the hybrids descended from *Hosta plantaginea* also require bright conditions and proper moisture for their best performance. The necessity for bright conditions also means that *H. plantaginea* and related cultivars are some of the most sun-tolerant of all hostas.

Another native of China, *Hosta ventricosa,* is not as particular in the garden. It has proven to be tolerant of a wide range of soil types and light levels. *Hosta ventricosa* blooms in July, producing bell-shaped, bright purple flowers in great profusion over a medium-large mound of shiny, dark green foliage. Copious numbers of seed pods develop following flowering. *Hosta ventricosa* is in the background of dozens of hybrids. Nearly all display shiny, dark green foliage and mid-summer purple flowers.

In its native Japan, *Hosta montana* grows in a wide variety of habitats from western Honshu into central Hokkaido. It can be found growing out of the cracks in rocks near waterfalls and rivers, at the edge of deep woods, and along roadsides. In the garden, *Hosta montana* exhibits a high level of adaptability, doing well in deeply shaded to brightly lit areas in average to rich soils. It has been widely used as a background or specimen plant as it forms a large, vase-shaped mound of foliage. The very pale lavender flowers of *Hosta montana* open in July on scapes held well above the foliage. Each flower is subtended by a bract, a characteristic shared by the many *Hosta montana* seedlings and sports.

The standards for blue-green leaf color, thick substance and an early blooming season in hostas come from *Hosta sieboldiana.* It forms a large-sized mound of slug-resistant, blue-green foliage and produces dense clusters of near-white flowers from mid-June into early July. At the same time, it has a notoriously slow growth rate, taking many years to reach a mature size. Providing extra fertilizer and moisture, though, speeds up its growth rate, reducing the time to maturity. The same is true for the closely related *Hosta* 'Tokudama'. It also displays the same deep blue coloration, thick substance, and early blooming season as *Hosta sieboldiana*.

Another slow grower is *Hosta nigrescens*, a Japanese species native to the northern part of Honshu Island. It distinguishes itself by a combination of a vase-shaped mound habit, tall scapes, thick substance, and rich blue cupped leaves. Most of these traits are also found in its seedlings. In the garden, *Hosta nigrescens* is another species that benefits from extra fertilizer and moisture.

Hosta yingeri is a relatively recent discovery, being identified in the late 1980s. Due to propagation by tissue culture in the 1990s, it quickly became popular amongst hosta collectors, home gardeners, and breeders alike. *Hosta yingeri* is distinguished by its shiny, dark green foliage and narrow-petaled, purple flowers in late summer, characteristics it usually transmits to its seedlings. In the garden it has proven to be of easy culture, forming a medium-sized, mature clump in three to four years.

Two species, *Hosta capitata* and *Hosta nakaiana*, are native to both Korea and Japan. They have much in common in terms of ease of culture and plant habit, and both have been the source of numerous hybrids. *Hosta capitata* produces an attractive cluster of purple flowers in early summer and can be used as a ground cover or specimen plant in the garden. *Hosta nakaiana* forms a much denser mound of foliage than *H. capitata*, and blooms at about the same time.

Found only on Cheju Island off the coast of South Korea, *Hosta venusta* ranks as one of the best mini-sized hosta species. Despite exhibiting diversity in the wild according to veteran plant explorers, a single, very attractive clone has been grown in Europe and North America for more than a half-century. It becomes a rounded, dense mound at maturity and produces purple flowers on scapes that are in neat proportion to the foliage. It makes a great edger in the garden and can also be grown in containers.

Hosta hypoleuca rates high on the distinction scale among *Hosta* species. In nature it grows on rocky cliffs in only a few sites in Japan. It produces just one leaf per division and displays a glaucous bloom on the leaf underside. Lavender flowers open in late July on long, recumbent scapes. With such unusual behavior, it comes as somewhat of a surprise that *Hosta hypoleuca* does not require special conditions in the garden. Also, hybridizers have found tremendous benefit from incorporating *Hosta hypoleuca* into their breeding programs.

In this book, Reggie Millette emphasizes the importance of hosta species by categorizing cultivars on the basis of their species background. This unique method to the study of hostas lays the groundwork for growing hostas successfully and understanding what they have to offer. Further, he reviews disease problems, discusses the most popular hostas, and focuses on miniature hostas. All in all, his approach will benefit all levels of hosta enthusiasts, from the home gardener to the sophisticated aficionado.

Mark R. Zilis
December 20th, 2012

Note: Mark Zilis is the owner of Q & Z Nursery in Rochelle, Illinois and the author of two books, *The Hosta Handbook* and *The Hostapedia.* Mark has introduced over 300 hostas to the nursery trade including such popular cultivars as 'Sugar and Cream', 'Great Arrival', 'Pineapple Upsidedown Cake', and 'Victory'. In 2001 the American Hosta Society awarded him the Alex J. Summers Distinguished Merit Award, its highest individual honor. Mark has studied hostas in gardens throughout the United States and explored the native habitat of several species during five trips to Japan.

Mark Zilis is considered today as one of the most known hosta hybridizer in North America and he is the author of the book *The Hostapedia.*

MY PASSION WITH HOSTAS

It has been a long time since 1995, the year my doctor, Dr Benrihmo, told me I was in a *burn out* situation.

So many things learned since my first flower bed.

So many hostas discovered since my neigbour gave me my first hosta, *Hosta lancifolia*, a green common hosta. And right after I got *Hosta undulata*.

Hosta lancifolia

Hosta undulata

This was the very beginning of a new collection and a great new passion.

I am the founder of the first association of hosta lovers in the Province of Quebec, in Canada. This society was established in April 2000 and still exists in 2013. Due to all its activities — meetings, plant exchanges, newsletters, garden visits — Q.H.H.S. offers everything to satisfy the local amateurs.

In 2004, Montreal botanical Garden was designated as an official Display garden by the American Hosta Society. With more than 500 hostas, it is a must within your garden tours.

Later in 2003, always involved with the genus hosta, attending most of the annual conventions of the American Hosta Society, visiting hundreds of specialized gardens, I wrote and was published the first book on hostas in French. Since then, 16,000 copies were sold and now out of print.

Today my collection is still expanding. I keep going. Now I have more than 700 different registered hostas. And rest assured that as you read these lines, I will have added more.

Chapter 1

Knowing about hosta culture, forms, flowers, sizes

Important terms:

Sport: an individual arising from the result of a mutation or chimera rearrangement that is genotypically or phenotypically different from the original individual.

Seedling: plant of seed origin resulting from the fusion of an egg and sperm (pollen).

Streaked: having a variegation pattern that has many non-connected light and dark variegated sections usually in streaks running nearly parallel to the midrib, an unstable form of variegation.

Note 1: When we give certain details on cultivars, we will give you these descriptions:

1: W.G. Schmid (ex Maekawa) (ONI) AHS, K. Walek (R) 2009
This information is from the hosta register kept by the registrar of the American Hosta Society. We will give you the name of the person who discovered this hosta and the name of the person who registered it. The year of registration also appears.

2: [Hosta 'Summer Fragrance' x Hosta 'Fascination']
Here are the names of the parents used to obtain this result.

Note 2: All the dimensions are always given by the hybridizer. The size of hosta plants varies not only with age, but also with soil and climatic conditions all are factors that could have an effect on a plants maturity, so the American Hosta Society determines the categories.

We consider now the average size for each cultivar at maturity as follows.

There are now five sizes:

Miniature hostas with 4 to 8 inches (10 to 20 cm)
Small hostas with 8 to 15 inches (20 to 40 cm)
Medium hostas with 15 to 24 inches (40 to 60 cm)
Large hostas with 24 to 35 inches (60 to 90 cm)
Very large hostas with more than 35 inches (90 cm)

Hosta of the year

With the increasing number of hosta cultivars being introduced each year it is increasingly difficult for nurserymen and gardeners to choose just the right hostas for their sales areas and gardens. Therefore the American Hosta Growers Association established the AHGA Hosta of the Year in 1996. It is selected by a vote of AHGA members. Award winners are hostas that are good garden plants in all regions of the country, are widely available and in sufficient supply and retails for about $15.00 in the year of selection. This is an insurance of success in growing your hostas.

Here are the selections since the beginning:

2014: 'Abiqua Drinking Gourd'
2013: 'Rainforest Sunrise'
2012: 'Liberty'
2011: 'Praying Hands'
2010: 'First Frost'
2009: 'Earth Angel'
2008: 'Blue Mouse Ears'
2007: 'Paradigm'
2006: 'Stained Glass'
2005: 'Striptease'
2004: 'Sum and Substance'
2003: 'Regal Splendor'
2002: 'Guacamole'
2001: 'June'
2000: 'Sagae'
1999: 'Paul's Glory'
1998: 'Fragrant Bouquet'
1997: 'Patriot'
1996: 'So Sweet'

The best way to grow your hostas

A gardener always looks for plants that are reliable, with a nice mix in foliage, and easy to cultivate and to maintain. Hostas fill for sure all these requests.

Sunny areas in our home gardens become shadier due to all the trees and shrubs getting larger with time. Most of the plants that we love have now growing problems and can't bloom as supposed to. In fact, the large majority of the plants need good sunny periods to perform well.

The shade loving hostas became rapidly the most sold shade plants in North America. Because the majority of them have large leaves,

they are able to let the water evaporate. The large leaves will also absorb the chlorophyll that gets thru tall tree foliage during the shorter periods of sun. In fact, they have the ideal conditions to perform where there are fresher temperatures and less light.

Hostas are plants with different sizes of foliage, offering all sorts of colors and different forms and textures. If you want plants being nice from spring till the first frosts in autumn, able to stop most of the weeds, able to grow at the same place for many years, and do not need to be divided, transplanted or fertilized often, and that are nicer with time… your choice is very limited. With hostas, you will have a choice between many thousands of different plants.

There are also giant hostas and some very small ones.

Walk thru old gardens and observe the flower beds. Try to find plants that are there since the birth of the garden and you will realize the list is short, especially if it is a very shady area. However it is not unusual to see a very big hosta. The older it is, the nicer it is.

There are some blues (*Hosta* 'Abiqua Blue Crinkles'), some greens (*Hosta* 'Irish Hills')

There are some yellows (*Hosta* 'Sun Power'), some variegated (*Hosta* 'American Dream')

There are some with corrugated leaves (*Hosta* 'American Misty'), some smooth leaves (*Hosta* 'Moonlight Sonata')

Some are very bizarre (*Hosta* 'Second Coming')

Even all white in spring (*Hosta* 'White Feather')

There are some very large ones (*Hosta* 'Montreal Blue Ice Storm'), some very small (*Hosta* 'Pixie Vamp')

And some very very large (*Hosta* 'Empress Wu')

The numerous choices in colors are a great asset for the gardeners. A young hosta plant often will have different colors as it will have at maturity. Margins, for example, will take few years to get their full colors. The majority of the books, catalogs and websites will show pictures of the mature plants. However, when you buy a young plant, you will notice many differences in forms and colors, which are perfectly normal.

Your hostas will take few years to become mature, often up to five or seven years. When you are planting your hostas always remember the mature size and plant them accordingly, especially if you don't want to move them in a few years.

Hostas don't need to be divided. But if you have the intention to do it, try not to do it before the spring of the third year, because it seems that the new plant explodes at that time. The division should be done, if possible, leaving the mother plant in place. Dig up one side of the plant and with a sharp knife, cut the crown with roots and detach it with your hands. Cover the hole left by the removed section of the plant with new soil and a few days later nothing will show.

You can also dig out the whole plant and remove all the soil around the roots. You will notice that the roots are all mixed up like spaghetti. If you sort them out you will be able to detach and divide the plant without hurting it. Washing in a lot of water or with the garden hose will help.

Never divide a plant you have just bought because it is never at its best. If you divide it then, the hosta plant might die after this division.

Hostas, with leaves showing a great amount of yellow, cream or white will need more time to mature due to less possibility to make chlorophyll. Because of this phenomenon, majority of the large hostas have green or blue colors and are most of the time uniform color. However don't think that miniature or small hostas, green or blue, are stunted plant. Generally, small hostas grow slowly.

The soil remains one of most important factor in the cultivation of hostas. The best soil must be loose, to retain the nutritive elements and water. The less recommended soils are the beach sands and the clay soil, the first will let water and air go right trough and the second won't let water and air go through. Hostas will tolerate salty water and winds, but won't resist in sterile soils. In this case add more peat moss. They prefer soils that are more acid than neutral, pH around 6. The hole for planting should be the same size of the root ball at maturity.

A good soil should be made of one third of planting soil, one third of coarse sand and a third of humus for at least 6 inches (15 cm). If you add organic matters regularly (compost, shredded leaves or well decomposed manure), your soil will increase its capacity to retain water and to keep the nutriment reserves, like nitrogen, potassium and phosphorus.

Many gardeners fertilize unnecessarily their hostas. There are no needs to give too much fertilizer to your hostas. Instead of fertilizer, add 4 inches (10 cm) of compost every two to three years. A good habit is to remove the dead flowers to stop the formation of seed pods. However if you leave them on the stems, you might have the good luck to see little new hostas in your flower beds. Water regularly in drought periods.

Weeds are almost non-existing under hostas, especially when they are mature. The big leaves will stop germination of weeds, by stopping light and water, two elements necessary to their germination. Don't grow spreading and aggressive ground covers like *aegopodium*, spreading *cotoneaster* or *pachysandra*. Be careful also with *vinca minor* that grows like weeds and will choke your hostas, especially the small ones. It is a good method to mulch your recently planted hostas in spring. Be careful not to choke the small ones with mulch, if you cover them totally.

The plants grown especially for their foliage like hostas will lose lots of water, even more if the soil is very humid. Hostas need just enough water to not wilt. Be more careful in drought periods. Here are a few advices about watering your hostas:

1. Water slowly (a vaporisation is better that any other type of watering), not letting the soil getting compacted, or encouraging erosion, and to leave the blue coating (*indumentum*) on your blue hostas.

2. Water in the morning, before the weather begins to warm up, allowing time to dry before night time (moment where the slugs are more actives).

If your hostas are not at their maximum, a small fertilization might help. A well balanced fertilizer will be ok. Hostas don't see any difference between natural and chemical fertilizers. However they like slow release fertilizers compare to the fast release chemical fertilizers.

Personally I use alfalfa pellets. This nitrogen source will favor nicer foliage. Apply in spring and mix it with the soil as soon the plants start coming out of the ground. Any other fertilization made after mid-July will favor the growth of new tender shoots, which will be vulnerable to slugs and diseases. New growths in late summer will not have time to harden before the first frosts and therefore will not survive the winter.

In general, hostas receiving more sun need a more humid soil. The more humid your soil is, the larger the leaves will be. Planted in shade, your hostas will have larger leaves, but with less blooms. The larger leaves will appear when your plant reaches maturity. Another thing to remember: a blue hosta planted in a sunny area will be greener.

The *indumentum* is the coating on blue leaves. That blue surface could be damaged by the sun. The surface might evaporate completely and

The same *hosta* 'Abiqua Drinking Gourd' at the left planted in a shady area and the one at the right is in a sunny area.

the foliage will become green. Blue hostas are at their best under a shady fresh area. Avoid rubbing the blue leaves, removing this blue coating and you get a green hosta.

A yellow hosta like 'Abiqua Recluse' will be greener planted in a shady area. (Right)

A hosta planted in a sunny area should produce more new growths, more flowers, but the leaves will be smaller.

It is possible that your hosta doesn't like the place where it is planted. Don't be afraid to transplant it somewhere else. Transplanting hostas is quite easy, especially during spring when the ground is not too warm. However you can do this anytime during the season, but wait for a cloudy day. Remove a large clump and don't break the small roots feeding the plant. If you transplant a hosta in the middle of summer in full sun, cut the leaves and water well. The new growths will be in proportion with the new root system.

Hostas growing in pots during the summer (in your landscaping) will often have roots growing outside the pots. You have to remove them before planting in the ground. It is also better to use new soil when you transplant your potted hosta. Dead or damaged roots should be removed. Water well.

When autumn arrives, just before winter, we recommand to cut the hosta foliage, especially if the autumn is very rainy. This procedure results in removing hiding places for the slugs and their eggs to survive.

Hostas are perennial plants. They should be planted to survive winter temperatures. Another method to keep them during winter consits in laying your pots sideways and covering them with a geotextile liner. A hosta in a pot fears frost and thaws.

If your soil is too compact, a brown margin will develop on the leaves, especially during drought periods. Improve your soil with more compost, not only near the base of the plant, but all around.

Watering with soaking hoses or overhead sprinklers is always better than a conventional watering.

Yellow hostas love light and heat and will light up shady areas.

Hostas need a dormant period to perform at their best. This dormant period should be around 700 hours (one month) with a temperature around 40 °F (5°C).
Some hostas like *Hosta* 'Elegans' take nice fall colors when autumn arrives.

How to plant hostas
Planting methods

1- Dig a hole at least twice the size of your pot.

2- When you take the plant out of the pot, unfold the roots, especially if the plant has been in the pot for a long period. Remove as much soil as possible. (washing helps)

3- Place your plant on top of at least 4 inches (10 cm) of your planting soil.

4- Add your fertilizer.

5- Fill with your planting soil. Make sure there are no air pockets in the ground.

6- Pack up some planting soil around your plant to prevent water from running away.

7- Install a plant tag to identify the new plant.

8- Water well with a watering can or a hose with a shower head. This type of watering will allow your plant to retain the water instead of soaking.

9- Cover with mulch to keep the ground humid.
(Personnally I use coconut mulch).

If you plant plugs (very small plants) the planting method is the same, but make sure you disturb the roots especially if they are too compact.

Flowers on hostas

I don't think it is necessary to explain in detail the morphology of the flowers.

The hosta flowers are variable in sizes. Forms can also vary. The majority of the hosta flowers are simple. But sometimes we have double flowers (*H. plantaginea* 'Aphrodite').

The flower colors range from dark purple to white, and all sorts of lavenders. Most of the hosta flowers are not fragrant. But in the last few years, hybridizers succeeded in the creation of new cultivars with very fragrant flowers that are now in our gardens.

Concerning the flowering periods, some early bloomers will show at the very beginning of the season, while some will bloom very late in autumn.

Many persons consider the flowers of hostas very insignificant. If you don't like them, just remove them. This will not harm the plant, but it should produce more roots.

However, if you plant hostas bearing fragrant flowers, plant them along a path, near a patio or even better under a window to take advantage of the perfume.

Chapter 2

Insects, diseases and virus HVX

Note: A part of this text was printed in my first book on Hostas by Les Éditions de l'Homme.

Hostas are known normally as no problem plants and almost no diseases.

The main enemies of hostas: slugs and/or snails

Only the slugs and/or the snails have a negative effect on the culture of hostas.

Some leaves on your hostas have irregular holes in the middle of the leaves and the holes are not necessarily round, sometimes quite long and look like lacerations. You will see the presence of slugs even before you see the damages on the leaves: look for the presence of mucus on the surface of the leaves. Taken at the beginning, the problem is easier to solve.

Damages by slugs could have different shapes

A slug is a mollusk same as oysters or mussels. A slug is different from a snail because, it doesn't have a shell to protect it. Slugs lay quantities of translucent eggs, like pearls, hidden under debris or buried in the ground. A slug can lay between 20 and 100 eggs, many times during the season, and each slug will approximately take two years to mature. A slug can live up to six years. The grey slug, *Peroceras reticulatum*, is the one we find generally in our flower beds. The grey slug measure around one inch in length. In the last few years the black slug appeared in some regions. This one is much larger. The slug is **hermaphrodite,** which means it has the two reproductive systems (both male and female) on each individual. Slugs have two pairs of tentacles on their forehead. The one on top allows them to see, while the one under allow them to smell. Under the body, there is a gland producing mucus. This foam stays until morning, and then becomes solid, reflecting the sun as it dries. This mucus allows the slug to keep its body always humid. Slugs have sharp teeth that can chew the middle of the leaves. Sometimes slugs can chew the rim of the leaves but most of the time, this damage is done by chewing insects.

Slugs start crawling and feeding when they are between one to three weeks old. It will take between two months to one year to become adult. Slugs can eat up to 30 to 40 times their weight in one day. They get older during winter and lay their first eggs in spring. The new adults are present in the gardens when the second generation arrives.

Damages on plants by slugs are only cosmetic. However if there is a big infestation, that can cause a stress to the plant and can even kill it in extreme cases.

A slug at work

How to solve the problem?

Gardeners and specialists have all kinds of remedies to eliminate or reduce damages to the plants by slugs.

First you can prepare your garden to make it less inviting for the slugs. Then try to attract natural predators. (birds, frogs, toads).

One thing important to know: slugs are always present in a garden. If you don't see them on your hostas, they might be on your tomatoes or lettuce. Don't leave a slug living. If you catch one in your garden, kill it. As soon you see one slug, take the best way for you to get rid of them, but act!

There are many recipes, not always ecologic, to eliminate the slugs. There are biologic ways, repression tools and finally the chemical weapons.

I prefer using the non toxic manners than all the poisons that could be detrimental for the environment.

Good preparation of the flower beds

You will notice that the slugs like hostas but don't like the acid substance. However hostas like acid soils. So make sure that your soil is more acid and often add peat moss and compost.

Slugs prefer humid sites and will hide under weeds. So don't let them have all kinds of easy places to hide: remove all weeds, mow your lawn very short, don't leave any debris on the ground (rocks, pieces of wood, bricks, containers, etc.).

In your flower beds, make sure there is a good aeration between the plants.

Use good mulch. Pine needles or crushed egg shells might fight back the slugs, either with the smell, resin or sharp surfaces. Gardeners had success by adding up to 2 to 3 inches (50 to 75 cm) of pine needles. These needles don't change pH in a significant way. But be very careful, some types of mulch will give slugs ideal places to hide.

Pine needles: a good way to eliminate damages caused by slugs

Flower bed maintenance

Clean your flower beds in spring. Rake the dirt deeply. Doing so, you will eliminate a large number of dormant slugs and quantities of eggs. Do this work on a sunny day. We all know that the slugs don't like sun. Wait when your soil is dry to do the work.

Remove the lower leaves as much as possible and don't let the leaves touching the ground. Stake the companion plants to allow the soil to be dry.

Make sure your flower beds have good drainage. Slugs are looking for humid soil to lay their eggs.

When you water your flower beds, do it early in the morning. Soil will have time to dry before night time. Slugs like humid and cool soils.

When choosing your hosta cultivars, select the slug resistant ones. Make sure also that your companion plants or trees are slug resistant.

Cultivation in raised flower beds or in containers is also less favorable to slugs.

Many recipes were tested to get rid of the slugs. Beer, wood ash or egg shells, all these recipes deserved to be tried, but you have to vary often. The best method to eliminate slugs remains the **hand picking**. At night when the sun is down, in the dark, you tour your garden with a flashlight. Put the slugs you find in a large bowl, full of soapy water. Of course you will collect only the one you can see, but it is the most efficient way to reduce their population.

If all these above methods don't interest to you, you still can grow and plant hostas. Slugs love hosta leaves because most of them have thin and tender leaves for the small slug teeth. Hostas with heavy substance leaves are slug resistant. Among all hosta cultivars, many are known to be slug resistant.

Another enemy for hostas

Deers

Deers like hosta leaves. In this case, not much can be done, except a 10 to 12 feet (3 meters) fence. It is the best solution, but quite expensive. It also seems that pieces of tissue soaked with perfume and installed here in there in the garden will chase these nice animals. It seems that the smell reminds them of human beings.

Some gardeners told me that a mixture of one egg in one or two liters of water, sprayed on hosta foliage might also keep the deers away.

Avoid a nice cedar hedge around your garden, if not protected. Deers like also these conifers.

Leaves bitten by deers

There are also rabbits and hares, moles and voles. For them, you need traps to try solving this problem.

Cutworms, ladybirds and grasshoppers are also insects to check.

The importance to know hosta virus HVX

Hosta virus X is now very important in the hosta world. So the explanations to understand this phenomenon have now become a necessity and we must continue to do everything to eliminate this problem from our gardens and nurseries.

The first infected hostas came from Japan. HVX was identified in United States in 1996 by Dr Lockhart from Minnesota University but this virus was seen on hostas as far as the 1980's. Because this virus is part of the potex family, it was given the name virus X.

Doctor Lockhart developed the ELISA test to analyze this virus. ELISA: Enzyme-Linked Immunosorbent Assay is used to detect various viruses, in many circumstances. During the tests, the chemical ingredients react and show the presence of antibodies associated to that kind of virus.

Our goal is to help all the ones interested with hostas:

- To know that virus HVX exists, that it can spread and is incurable.
- Learn how to recognize the symptoms of the virus.
- And learn how to stop that problem.

By showing you these facets, you will be checking more carefully, it will be easier to stop this virus from spreading and you will understand the importance to have healthy plants in our gardens and in the nurseries.

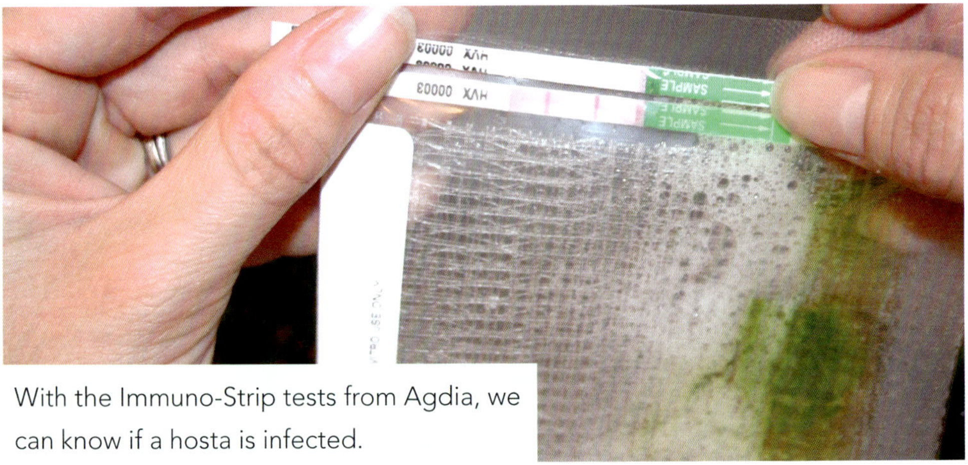

With the Immuno-Strip tests from Agdia, we can know if a hosta is infected.

Since 2006, we have less infected hostas on the shelves at the wholesalers. However we can always find some infected hostas in a nursery, because many of these plants were bought many years ago, when information was less available.

Even today there are many infected hostas in private gardens and the number is getting higher. We must educate the population on how to detect an infected plant and on how we can stop the transmission on other plants.

HVX like other viruses will infect the plant forever, so the plant will degrade in its look and health.

The cultivar won't look like it should be.

Hosta 'Katherine Lewis' healthy

Hosta 'Katherine Lewis' with the virus

Hosta 'Katherine Lewis' with the virus

Hosta 'Gold Standard' healthy

Hosta 'Gold Standard' with the virus

Hosta 'Gold Standard' with the virus

The plant might be deformed, have growing problems, and generally have an ugly look with spots, light or dark coloration, along the veins and the hosta a scaly appearance. Leaves have a flat look and their growing speed is slower. They are also smaller than the same healthy plant.

We can also see spots on leaves, lightly translucent and sometimes they have unusual twisted leaves.

HVX, like all viruses, can be transmitted to other hostas. We don't know yet if this virus can be transmitted to other types of plants. HVX, like many viruses, is incurable.

In fact it is always important to buy your hostas thru a grower knowing about this problem or from someone that you are certain has hostas that are virus free.

Yes! You can help the destruction of hostas by virus X

Inform your friends and neighbors if you see an infected hosta in their garden.

Show them these informations and the photos if they are not convinced.

HVX is still quite recent and it is possible that certain nursery people were not advised. If you see plants that look infected talk to the nursery owner or the manager. Tell him about the virus.

How the virus is spreading?

HVX is the type of virus spreading in a mechanical way that means it is spread by the human.

These kinds of damages appear on a plant and transmitted to another healthy plant. Here are some examples:

1- Cutting a leaf, a petiole, a flower stem on an infected hosta, where the infected sap is spread on a healthy plant using the same tool to cut leaves, petioles or flower stems.

2- Cutting the crown of an infected plant followed by cutting the crown of another hosta, when dividing for example.

3- Digging out a plant and cutting the roots on an infected plant followed by cutting roots on a healthy plant. This can happen when weeding, hoeing or transplanting.

Never keep an infected plant. destroy it to eliminate all possible epidemics.

Have a garden in good health at all times

You must clean all your instruments before acting between each hosta plant. Clean the shovels, pruners, knives, or scissors that were in contact with any hostas. Wash your hands after cutting your hostas, or after dividing, or having touched any cut parts of the plants. Make sure to clean your tools after use. In a case where it was an infected hosta, it is more important to clean your tools with a towel to make sure you remove any infected liquid. Disinfect your tools with alcohol or with cloride. It is the liquid... the sap which transmit the virus. So if you clean your tools, you reduce the chances of infecting other plants. It seems that you can neutralize the virus if you clean your tools and / or let them soak in these solutions.

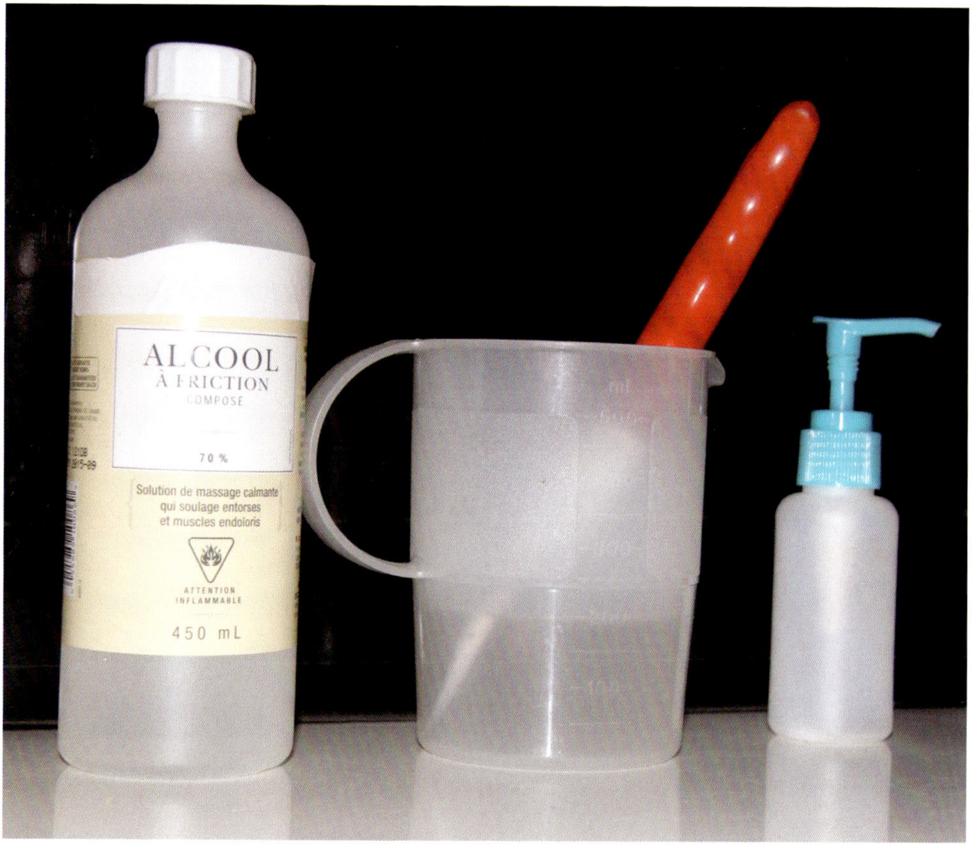

By leaving your pruner in a water mixture with alcohol, you can stop the spreading of the virus from a plant to another when you reuse the same tool.

Chapter 3

Hybridizing hostas

It is very easy to hybridize to obtain a new hosta. But to be good enough to be registered, it must be very exceptional.

On each flower there is a male organ (stamens and pollen) and a female organ (pistil).

When your hostas are blooming, you will see a yellow powder (pollen) on the stamens. On the same flower you will find a longer part call the pistil.

Early in the morning on a warm day, take a stamen with its pollen. Remove gently a stamen. Then spread the yellow pollen on the sticky end of the pistil of another flower on a different hosta.

To avoid any insects to transfer pollen on the same flower, remove all petals of the flower and the unused stamens.

You can also store pollen in a small sealed container placed in refrigerator for a future use.

About six weeks after pollination, seed pods full of seeds will appear on the flower stem of your hosta plant. When the seed pod is brown and starts to open, it is time to collect the seeds.

Hosta with many seed pods

You are now ready to sow the seeds in the soil or in small containers. It will take only a few years, four to six years before you know if the new hostas will have the same characteristics that you wanted.

As you can see hybridizing is quite easy.

The difficult part is to discover the best parents that will give the expected results. In all that, you need luck. In the same seed pod you will find a great diversity. Some, when getting older will take some strong traits of one parent while another will take the strong traits from the other parent. The result is always a surprise. You will have to wait many years to make sure it stays stable.

Every year when new plants come out at shows, we can see hostas with red stems always winning the responses from judges. So, many

In a seed pod, the seedlings won't be identical plants.

hybridizers are searching for parents able to transmit these characteristics in their breeding program.

Each year a hosta is honored for its performance in the gardens. In 2012, Olga Petryszyn received the award for *Hosta* 'Niagara Falls'. Here is the acceptation speech when accepting the Benedict Medal by Olga Petryszyn for *Hosta* 'Niagara Falls'.

1-M-15
GREEN
SOCIETY
REGION 1
SEEDLING
AND SPORT
COMPETITION
BEST OF
SECTION

The allocution resumes all the steps that favored the creation of the *Hosta* 'Niagara Falls'. Thanks to Olga for allowing me to publish her speech.

Hello, my name is *Hosta* 'Niagara Falls'. If I may, I'd like to tell you a story my parents told me about how I came to be. Once upon a time on a 25

acre homestead in Indiana lived two very talented men. They partnered to create a lovely home and gardens in the country. Their names were Bill Brincka and Basil Cross. In about their 30th year of gardening, Bill had the idea to create another garden room in the woodland.

Bill and Basil had a pretty friend that lived a mile away. Her name is Olga. Bill mentored her as an art student and taught her about gardening. One day Olga came by to talk about the new garden project. There Olga saw the ropes strung between the trees outlining what Bill called the Hosta Garden. In about three months some trees were cleared, a path was put in and about 20 different hostas were planted. Now Olga knew about 3 or 4 kinds of hostas and was very surprised by the varieties. She just loved them. Bill suggested she try hybridizing them and create new ones. Bill would purchase any new available hostas and he and Basil would plant them. Olga could come by and play with them.

Early in the morning, July 4th, 1986, Independence Day, Olga sat in the hosta garden examining the qualities of each and wrote down her plan. She loved mom's size and cascading quality. Her name is *Hosta montana* 'Praeflorens'. Since then her name has been changed to *Hosta montana macrophylla*. Then the rippled edges of another hosta really caught her eye. He looked so playful. That's my dad, *Hosta* 'Sea Drift'. That morning Olga performed the nuptials, they were married and I was conceived. I grew to a healthy seed along with many other seeds, snug in a pod hanging on to mom. From that point in my life I can just start remembering my past.

Fall approached and Olga plucked us from mom and pushed us out of our pods. Then she placed us in an envelope and into a cold refrigerator. This frightened us but then we just went to sleep. On February 2nd, 1987, Groundhogs Day, all the seeds were taken out of the frig and spread out onto wet warm soils in a tray. This felt great except for the bright lights left on 24 hours a day. In about ten days something strange happened. I swelled and grew a tail that went into the ground and a head of sprouts that turned to leaves, stretching for the light. I was crowded by my brothers and sisters that were doing the same thing. The crowding was so tight that only the strongest grew big. Then on July 4th, Independence Day 1987, a year since I was conceived, that lady Olga took us outside. The air was fresh and it was a dark night. The sounds were wonderful. The next day we saw sunshine for the first time. Olga scooped us up in her hands and with her fingers starting taking us apart. It felt wonderful to have my roots untangled. By Olga's side was a bucket and in it were about 80% of my siblings lying is a heap. Then that heap was thrown on a mulch pile. Who is this woman? Then Olga took me and the rest of my siblings, 17 in all, and planted us in rows. We were watered and fed and treated so well. When winter came, we went to sleep and the following spring we grew bigger. The next year we grew bigger yet.

When I was three, Olga dug us all up. She tossed everyone in a pile except me and my sister. I felt lucky but was not sure I liked this lady anymore. But then my sister and I were planted in 2 separate raised beds and treated like royalty. At the age of 6, Olga's friends Bill and Basil came to visit Olga's garden and see how the various hosta babies were doing. They stood in front of me and said 'WHO IS THAT?' The smiles and attention made me feel proud and I held my leaves up as high as I could to show off. I think I actually shivered a bit. Olga told them I was the child of *Hosta montana macrophylla* and *Hosta* 'Sea Drift'. Bill said "That Hosta is going to be famous some day and it will put your name on the map" Olga said I needed a name. Basil suggested *Hosta* 'Niagara Falls' for the cascading ripples. We all thought it was perfect. Olga put my name on a label and placed it in front of me. Later I was photographed many times and measured up. Then I was put in the records for registration with the American Hosta Society. My sister was given to a friend of Olga's. She grew bigger than me but I had a darker color, better ripples and deeper veins. She is known as *Hosta* 'Niagara's Sister'. I was told she gets a lot of ooohs and aaahhs.

Mom and Dad told me I was sold in the early 90's to collectors for $100.00 a division. How about that, I contributed to the economy! Who knows, maybe I'll end up in the Millennium Seed Bank in West Sussex!

Over the years, I have been spread throughout the United States, Canada and Europe. Olga and her husband Lew even visited me at the Tower of London where I grow with other hostas. My parents said that I have produced many Children, Grandchildren, Great grandchildren, Great Great grandchildren, nieces, nephews, cousins....you name it. All of them sharing some of my fine qualities that I have been so happy to share.

Some of my genes helped create: *Hosta* 'Hoosier Dome', *Hosta* 'Yosemite Valley', *Hosta* 'Continental Divide', *Hosta* 'Coal Miner', *Hosta* 'Dune Boy', *Hosta* 'Warpaint', *Hosta* 'Bridal Falls' and many other seedlings waiting to grow up.

I am only a green hosta who was lucky enough to be created and chosen. Who would have thought that some day I'd be famous and live into infinity? I am so thankful to be admired and have my traits used for other creations. Kudos to Bill, Basil and Olga for giving me life. Olga is not so bad after all.

Marco Franzen from Netherland is a great hosta grower in Europe.

Chapter 4

Best way to succeed with your hostas

We all know that hostas are the most sold perennials for shade in North America.

There are more than 40 species found in the world of hostas. Each of these species has their particular way of living in the wild nature.

However we know that many hostas can tolerate sunny periods and their foliage won't be too damaged. As it is very difficult to determine which of the 9000 hostas will tolerate sun, we will analyse some species to know their growing habits in their original countries. By analyzing each of these, we should be able to determine how they were living: type of soil, sunny periods, growth speed, sizes, etc.

After that we will find all their sports or mutations produce by the plant and all their seedlings, having parents from members of the species. We had to make sure these hostas have the growing habits of the species. It would be logic to obtain good results with these hostas if we give them the same type of habitat.

Thanks to Mark Zilis, author and well known hosta specialist for helping me verifying and endorse the lists.

Hosta plantaginea species

This species from the hosta world is by far the one that tolerates the most heat, sun and humidity as it origins in the most Southern Country of its world distribution. Most of the hosta species were found in Korea or in Japan while *Hosta plantaginea* origins in Eastern China, near Beijing or Shanghai where the weather can be very hot and very humid.

Because of its heat and humidity tolerance, this species produces

large white very fragrant flowers and favors new leaf growth of leaves during the season.

Hostas from *plantaginea* type emerge very early in spring. However, it is possible in your garden, if located in colder winter climates, where frost may still come late in spring that you could have to protect them with mulch.

The white flowers appear late in the season, that's why they were called 'August lily'. Usually it is the summer period where weather is warmer and there is more rain falls that are at their best. Hostas *plantaginea* and its progeny are the only hostas producing flowers that open late in the afternoon. The perfume is remarkable. We all should give a preference in planting them in places where we always sit in the garden, near a pathway, or near a window, just to appreciate the nice scent. If there is no rain in August, **WE MUST WATER THEM!**

This species is always very popular with the hybridizers. It gives a lot of **sports (mutations)**, sometimes very different from the mother plant. Note also that a very large majority of these new mutations come from tissue culture.

By analyzing all the different plants from that group and giving them almost the same conditions as where they were found, we should have a better chance of success with this type of hostas.

Plants being sports of *Hosta plantaginea*

Hosta 'Aphrodite'

W.G. Schmid (ex Maekawa) (ONI) AHS, K. Walek (R) 2009

A large pale green hosta; leaves are shiny on top, shiny underneath and smooth giving double flowers. Must be watered often when very hot periods if you want to see the buds in bloom. In the cooler climates, the buds might abort.

Hosta 'Avocado'

J. & B. Diesen (ONIR) 1998

[Sport of *Hosta* 'Guacamole']

A large yellow gold hosta with a green margin. Leaves are smooth, very shiny on top and very shiny underneath. Looks like *Hosta* 'Guacamole', but the margin is wider.

Hosta 'Fragrant Dream'

Marco Fransen NR

[Sport of *Hosta* 'Fragrant Bouquet']

This hosta in not registered and was discovered in Netherlands. A medium fast growing hosta with leaves being apple green, streaked and slightly shiny. Leaves have an irregular yellow margin.

Hosta 'Fried Bananas'

R. Solberg (OR) 1994

[Sport of *Hosta* 'Guacamole']

A large yellow gold hosta with good growing speed. Leaves are smooth, very shiny on top and very shiny underneath. Looks like *Hosta* 'Fragrant Gold' and *Hosta* 'Tortilla Chip'.

Hosta 'Fried Green Tomatoes'

R. Solberg (OR) 1995

[Sport of *Hosta* 'Guacamole']

A green medium size hosta. Leaves are smooth, very shiny on top and very shiny underneath.

Hosta 'Guacamole'

R. Solberg (OR) 1994

[Sport of *Hosta* 'Fragrant Bouquet']

A medium size hosta. Leaves are yellow gold with a dark green margin. Leaves are smooth, very shiny on top and very shiny underneath. Looks like *Hosta* 'Avocado' and should be in a garden.

Hosta 'Miss Saigon'

H. Hansen (ON) H. Hansen & Shady Oaks Nursery (IR) 1999

[Sport of *Hosta* 'Fragrant Bouquet']

A medium size hosta. Leaves are dark green with a creamy white margin. Leaves are dull on top and underneath. Looks like *Hosta* 'Fragrant Dream', *Hosta* 'Fragrant King' and *Hosta* 'Fragrant Surprise'.

Hosta 'Royal Super'

Walters Gardens, Inc. (ONIR) 2000

[Sport of *Hosta* 'Royal Standard']

A medium size hosta. Leaves are green with a creamy white margin. Leaves are slightly shiny on top and underneath. Very nice foliage.

Hosta 'Sugar and Cream'

M. Zilis (R) 1984

[Sport of *Hosta* 'Honeybells']

A fast growing large hosta. Leaves are dark green with creamy white margin with slight undulations. Looks like *Hosta* 'So Sweet'.

Hosta 'Sweet Innocence'

Shady Oaks Nursery, NR

[Sport of *Hosta 'Fragrant Bouquet'*]

A medium size non registered hosta. Leaves are light apple green with a yellow cream margin. Margin becomes white along the season. Leaves have good substance

Hosta 'Venus'

Walters Gardens, Inc. (ONI) Walters Gardens, Inc. & C. Falstad (R) 1993

[Sport of *Hosta plantaginea* 'Aphrodite']

A large green hosta. This hosta has white double flowers. Leaves are shiny on top and underneath. Must be watered often during the season.

***Hosta plantaginea* produced many other sports:**

- *Hosta* 'Athena'
- *Hosta* 'Color Parade'
- *Hosta* 'Fragrant Flame'
- *Hosta* 'Fragrant Light'
- *Hosta* 'Heaven Scent'
- *Hosta* 'Hoosier Harmony'
- *Hosta* 'Julia Gaede'
- *Hosta* 'Ming Treasure'
- *Hosta* 'Marbled Bouquet'
- *Hosta* 'Parthenon'
- *Hosta* 'Paul Aden'
- *Hosta* 'Poseidon'
- *Hosta* 'Prairieland Memories'
- *Hosta* 'Redolent Nosegay'
- *Hosta* 'Royal Accolade'
- *Hosta* 'Royal Standard Special'
- *Hosta* 'Sun Bouquet'
- *Hosta* 'White Shoulders'
- *Hosta* 'Winter Bouquet'

This hosta was used often in hybridizing and results with many seedlings with the same habits of

Hosta plantaginea.

Hosta 'Austin Dickinson'

W. & E. Lachman (OR) 1992

[*Hosta* 'Resonance' × *Hosta plantaginea*]

A small hosta. Leaves are green with a pure white margin. Leaves are slightly shiny on top and dull underneath. Interesting in the garden all season long.

Hosta 'Emily Dickinson'

W. & E. Lachman (R) 1987

[Hybrid of *Hosta* 'Neat Splash' × *Hosta plantaginea*]

A fast growing small hosta. Leaves are green with a creamy white margin. This hosta offers good combination with margined foliage and fragrant flowers.

Hosta 'Flower Power'

Vaughn K NR

[*Hosta nigrescens* x *Hosta plantaginea*]

A non registered large grey green hosta. Leaves are slightly ruffled. This hosta offers a good combination with the characters of both parents.

Hosta 'Fragrant Bouquet'

P. Aden (R) 1982

[*Hosta* 'Summer Fragrance' x *Hosta* 'Fascination']

A fast growing small hosta. Leaves are apple green and streaked yellow and white. A very popular hosta due to its nice foliage and fragrancy.

Hosta 'Green Marmalade'

Clarence Owens 1987

[*Hosta* 'Neat Splash' x *Hosta plantaginea*]

A large hosta. The leaves are green, mottled and glossy on top. This hosta could be a star in a landscaping.

Hosta 'Honeybells'

A Cummings & AHS (R) 1986

[Parentage unknown, but we are sure it has the genes of *Hosta plantaginea*]

A fast growing large green hosta. Leaves have large undulations.

Hosta 'Invincible'

P. Aden (R) 1986

[*Hosta* 'Aden 314' × *Hosta* 'Aden 802']

A good growing medium size green hosta. The glossy foliage makes it very popular.

Hosta 'Iron Gate Glamour'

V. Sellers (R) 1981

[Hybrid of *Hosta plantaginea*]

A medium size hosta. Leaves are deep green and mottled, with a fine white margin.

Hosta 'Old Faithful'

W. Brincka and O. Petryszyn (R) 1991

[*Hosta plantaginea* × *Hosta* 'Regal Rhubarb']

A medium size green hosta with glossy leaves. When mature this hosta will have a vase shape.

Hosta 'Rippled Honey'

R. Benedict (ONI) P. and J. Ruh (R) 2005

[Parentage unknown, but we are sure it has the genes of *Hosta plantaginea*]

A fast growing little green hosta. Leaves are heavily rippled, slightly shiny on top and underneath. Leaves are moderately cupped and twisted. This hosta looks like *Hosta* 'Invincible'.

Hosta 'Royal Standard'

J. Grulleman and Wayside & AHS (R) 1986

[*Hosta plantaginea* × *Hosta sieboldiana*]

A fast growing large green hosta. Leaves have slight undulations and have a glossy sheen. This hosta was one of the first hosta sold in America, but always popular.

Hosta 'Snow Flakes'

F. Williams and AHS (R) 1986

[*Hosta sieboldii* 'Alba' × *Hosta plantaginea*]

A fast growing green hosta. This hosta can be used as a border or ground cover.

Hosta 'So Sweet'

P. Aden (R) 1986

[*Hosta* 'Fragrant Bouquet' × *Hosta* 'Aden 462']

A fast growing small hosta. The leaves are green with a white margin. Being popular it was name the first 'Hosta of the year' in 1996.

Hosta 'Summer Fragrance'

K. Vaughn (R) 1983

[*Hosta* 'Vaughn 73-2' × *Hosta plantaginea*]

A smal to medium size hosta. Leaves are deep green with a creamy white margin. This hosta was the first variegated hosta to be registered.

Hosta 'Sweet Sunshine'

B. Solberg 1997

[*Hosta* 'Sweet Susan x *Hosta* 'Tokudama']

A medium size yellow gold hosta. Leaves are slightly shiny on top and underneath. The foliage has the genes of the other parent *Hosta* 'Tokudama'.

Hosta 'Sweet Susan'

F. Williams and C. Williams (R) 1986

[*Hosta sieboldii* × *Hosta plantaginea*]

A fast growing small green hosta. Sun tolerant and the flowers open in the morning.

Hosta 'Sweetie'

P. Aden (R) 1988

[*Hosta* 'Fragrant Bouquet' × *Hosta* 'Fragrant Candelabra']

A fast growing small hosta. The leaves are yellow to bright chartreuse with a white margin. Leaves are also undulated. This hosta looks like *Hosta* 'Fragrant Bouquet'.

Hosta 'Warwick Essence'

L. G. Jones (ONIR) 1993

[*Hosta* 'Northern Halo' × *Hosta plantaginea*]

A large blue green hosta. Leaves are slightly shiny on top and underneath. One of the first hosta to bloom early in the season. This hosta looks like *Hosta* 'Flower Power'.

Many other hostas had *hosta plantaginea* as a parent:

- Hosta 'Atomic Mama'
- Hosta 'Bennie McRae'
- Hosta 'Black Mamba'
- Hosta 'Blushing Bride'
- Hosta 'Bob'
- Hosta 'Buckwheat Honey'
- Hosta 'Calamity Jane'
- Hosta 'Carrie'
- Hosta 'Chelsea Ore'
- Hosta 'Curly Top'
- Hosta 'Elp 2'
- Hosta 'Extasy'
- Hosta 'Flower Child'
- Hosta 'Forever Green'
- Hosta 'Fragrant Blue'
- Hosta 'Fragrant Gold'
- Hosta 'Gilda'
- Hosta 'Gold Piece'
- Hosta 'Iron Gate Bouquet'
- Hosta 'Iron Gate Delight'
- Hosta 'Iron Gate Supreme'
- Hosta 'Karenka'
- Hosta 'Lavender Lady'
- Hosta 'Long Tom'
- Hosta 'Mirage'
- Hosta 'Mistress Mabel'
- Hosta 'Pancakes'
- Hosta 'Precisely'
- Hosta 'Purple Profusion'
- Hosta 'Raleigh Remembrance'
- Hosta 'Seventh Heaven'
- Hosta 'Showtime'
- Hosta 'Sibling to Summer Fragrance'
- Hosta 'Slim Polly'
- Hosta 'Spit Shine'
- Hosta 'Statue of Liberty'
- Hosta 'Sugar Babe'
- Hosta 'Surfer Dude'
- Hosta 'Surprise'
- Hosta 'Sweet Marjorie'
- Hosta 'Tinker Bell'

To summarize, if we want to have success with hostas having *Hosta plantaginea* in their genes, it is preferable to plant them in sunny and busy areas to smell the fragrant flowers. These hostas prefer hot, sunny and humid areas. They can tolerate long hot periods.

These hostas will produce new leaves all season long.

Hosta montana species

This hosta species was discovered on Honshu Island in Japan by F. Maekawa in 1940. It was found on the edges of forests and in open meadows.

This vase shape hosta is large and sometimes very large, reaching sometime more than 3 feet (1 meter). The very large leaves are green, deeply veined and undulated. The plants emerge early in spring and are sensitive to frost if not protected. Leaves damages by frosts have a sloppy appearance. When summers are dry and very hot, leaves could burn. It is better to plant these hostas in shady areas.

Flowers of this species are light lavender. This plant is also very important for hybridizers.

A large green hosta.

Sports or mutations of *Hosta montana*:

Hosta 'Ebb Tide'

E. Elslager (ONIR) 1997

[Sport of *Hosta montana* 'Aureomarginata']

A very large hosta. Leaves are dark green with a yellow gold margin. Leaves have heavy ripples and are shiny on top and underneath. This hosta looks like *Hosta montana* 'Aureomarginata' but the margin is narrow.

Hosta montana 'Aureomarginata'

AHS (R) 1987

[Sport of *Hosta montana*]

A very large hosta. Leaves are dark green with a yellow green margin. Leaves have some undulations. This hosta is always very popular among the hosta lovers.

Hosta 'On Stage' (aka *Hosta* 'Choko Nishiki')

P. Aden (R) 1986

[Sport of *Hosta montana*]

A large hosta. Leaves are creamy white with a green margin. The center of the leaves will turn green during the season. One of the hostas that emerge very late in spring. This hosta can also bear the Japanese name *Hosta* 'Choko Nishiki'.

Hosta 'Praeflorens'

Maekawa - 1940 – NR

[Hosta discovered in Japan in 1940]

A giant vase shaped green hosta. This hosta was named *Hosta montana var. macrophylla* as believed to be a sub-species. Due to the thick leaves it is slug resistant.

There are few more mutations, hard to find, most of them from Japan.

We have today many cultivars that have been crossed with

Hosta montana

Note: Most of the plants from that species are quite large, it is possible that some of them would bear the other parent's genes.

Hosta 'Alice Gladden'

D. & J. Ward (ONIR) 1998

[*Hosta* 'White Vision' × *Hosta montana macrophylla*]

A medium size yellow gold hosta. Leaves are slightly shiny on top and dull underneath.

Hosta 'Big Boy'

L. Simpers (R) 1980

[Hybrid of *Hosta montana*]

A very large green hosta. Leaves have a slight rugosity and deeply veined. This hosta looks like *Hosta* 'King Michael'.

Hosta 'Dragon Wing'

W. Lefever (ONIR) 1996

[*Hosta montana* × *Hosta* 'Fortunei']

A large hosta. Leaves are medium green with a green margin. Leaves are very rippled. Leaves are shiny on top and underneath. This hosta looks like *Hosta montana*.

Hosta 'Frosted Jade'

L. Maroushek (R) 1978

[Hybrid of *Hosta montana* 'Krossa B-1']

A large hosta. Leaves are dark jade green with a creamy white margin. Leaves have undulations. This hosta is still popular. This hosta looks like *Hosta* 'Crispula' and *Hosta* 'Mountain Snow'.

Hosta 'Great Lakes Gold'

D. & J. Ward (ONIR) 2000

[*Hosta* 'White Vision' × *Hosta montana macrophylla*]

A large yellow gold hosta. Leaves are slightly rippled. Leaves are dull on top and dull underneath. This hosta looks like *Hosta* 'Choo Choo Train'.

Hosta 'King James'

G. Krossa (O) A. Summers (N) P. Ruh (R) 1992

[Hybrid of *Hosta montana*]

A large medium grey green hosta. Leaves are dull on top and satin sheen underneath. This hosta looks like *Hosta* 'King Michael'

Hosta 'King Michael'

G. Krossa (O) A. Summers (N) P. Ruh (R) 1992

[A select form of *Hosta montana*]

A large green hosta. Leaves are satiny sheen on top and glossy underneath. This hosta is often compared to *Hosta* 'King James'. But *Hosta* 'King Michael' is greener.

Hosta 'Lakeside Cha Cha'

M. Chastain (OR) 1994

[*Hosta* 'Fascination' × *Hosta montana*]

A large hosta. Leaves are light green with a creamy white margin. Leaves are slightly rippled, shiny on top and glaucous underneath. A very distinctive hosta with its fine margin. This hosta looks like *Hosta* 'Moonlight' and *Hosta* 'Sea Dream'.

Hosta 'Memories of Dorothy'

H. Gowen (OIR) H. Hansen (N) 2002

[(*Hosta* 'Dorothy Benedict' x *Hosta pycnophylla*) × *Hosta montana* 'Aureomarginata']

A medium size hosta. Leaves are medium green with a gold and cream margin. Leaves are lightly corrugated, slightly shiny on top and glaucous underneath.

Hosta 'Mikado'

P. Aden (R) 1982

[*Hosta montana* 'Aureomarginata' × *Hosta* 'Big Sam']

A very large green hosta. Leaves are slightly rippled. This hosta is often compared to *Hosta macrophylla*. It looks like *Hosta* 'Big Boy' and *Hosta* 'King Michael'.

Hosta 'Niagara Falls'

W. Brincka & O. Petryszyn (R) 1991

[*Hosta montana macrophylla* × *Hosta* 'Sea Drift']

A large dark green hosta. Leaves have small deep undulations. Leaves are glaucous on top. This hosta received the highest Awards.

Hosta 'Silk Kimono'

H. Gowen (O) H. Hansen (N) H. Hansen & H. Gowen & R. Benedict (I) H. Hansen & H. Gowen (R) 1999

[*Hosta* 'Dorothy Benedict' × *Hosta montana*]

A large hosta. Leaves are medium green to grey green with a creamy white margin. Leaves are dull on top and dull underneath.

Hosta 'Stingray'

R. Herold (ONIR) 2000

[*Hosta kikutti* × *Hosta montana*]

A large green hosta. Leaves are heavy rippled. Leaves are shiny on top and underneath. This hosta looks like *Hosta montana*, but the foliage have more undulations. Be careful, this hosta can be very large.

Hosta 'Yellow River'

E. Smith (O) P. Aden (I) P. Ruh (R) 1993

[Parentage unknown but we are sure it has genes from *Hosta montana*]

A large hosta. Leaves are medium green with a medium yellow marbled margin. This fast growing hosta, widely rippled, is very popular. This hosta looks like *Hosta* 'Abba Dabba Do'.

Many other seedlings have genes from *Hosta montana*.

- *Hosta* 'Amazing Grace'
- *Hosta* 'Arnold Black'
- *Hosta* 'Cascades'
- *Hosta* 'Chirifu Tochigi'
- *Hosta* 'Creech'
- *Hosta* 'Devon Tor'
- *Hosta* 'Elata'
- *Hosta* 'Emerald'
- *Hosta* 'Godzilla'
- *Hosta* 'Grand Canyon'
- *Hosta* 'Green Acres'
- *Hosta* 'Hida no hana'
- *Hosta* 'Jackpot'
- *Hosta* 'Kinkaku'
- *Hosta* 'Laura'
- *Hosta* 'Magic Mountain'
- *Hosta* 'Mesa Verde'
- *Hosta* 'Midnight Sun'
- *Hosta* 'Mountain Snow'
- *Hosta* 'Mount Everest'
- *Hosta* 'Mount Fuji'
- *Hosta* 'Night and Day'
- *Hosta* 'Outrageous'
- *Hosta* 'Pewter Frost'
- *Hosta* 'Prince of Wales'
- *Hosta* 'Satin Beauty'
- *Hosta* 'Smoothie'
- *Hosta* 'Stardust'
- *Hosta* 'The Hulk'
- *Hosta* 'Vanguard'

To summarize, if we want to have success with hostas having Hosta montana in their genes, it is preferable to plant them in shady areas. Most of these plants will be large or very large. Good plants for hybridizing.

Hosta sieboldiana species

This hosta species, the most widespread, was found on the west coast of Japan, a region where the climates are fresh, ideal for plants with a short growth period, and they like cool weathers. These plants emerge late in spring and bloom fast, set seed and prepare to go dormant for the winter. Generally, they form a dormant bud on the side of the crown. They also have only one set of leaves in spring. As it's their habit to support fresh temperatures, they can suffer easily and fall dormant if the season is too warm. Their dormant period will be activated when temperatures reach 95 F (35 C). In fact tissues get dormant and stay dormant for the rest of the season. The leaves don't die and don't disappear but can show signs of burns. It is the reason why hostas of the *sieboldiana* type are not sun tolerant. To avoid the dormancy due to great heat, keep watering the plant to give the roots a chance to cool off and grow. To avoid the problem you should also plant them in a shadier area.

The average of these large size plants is from 24" (60 cm) up to 60'' (1m50) in diameter when mature. Flowers are almost white and bloom from the end of June to the middle of July. Leaves are round, blue green, and have good substances and are undulated.

The real species are hard to find.

Plants having genes of *Hosta sieboldiana*

Hosta 'Big Daddy'

P. Aden (R) 1976

[Sport of *Hosta* 'Fortunei Robusta Variegated' but has signs of genes from *Hosta sieboldiana*]

A large slow growing blue green hosta. This hosta looks to *Hosta* 'Sea Lotus Leaf'.

Hosta 'Borwick Beauty'

G. & B. McBurnie & BHHS (R) 1988

[Sport of *Hosta sieboldiana*]

Large hosta. Leaves are light chartreuse with a blue green margin. As the majority of the hostas of this type, it will be easily touched by the sun burns early in the season. To avoid this problem, plant the hosta in a more shady area.

Hosta 'Frances Williams'

F. Williams & C. Williams (R) 1986

[Sport of *Hosta* 'Elegans']

A very large hosta. Leaves are blue green with chartreuse to yellow margin. As the majority of the hostas of the same type, it will burn easily. To minimize this problem plant it in a more shady area. You should also plant it in the back of a landscape to avoid questions about the appearance.

Leaves of this hosta always show signs of burns. This hosta looks like *Hosta* 'Olive Bailey Langdon' and *Hosta* 'Samurai'.

Hosta 'George Smith'

G. Smith (R) 1983

[Sport of *Hosta* 'Elegans']

A large size hosta. Leaves are golden lime with a blue green margin. This hosta looks like *Hosta* 'Borwick Beauty'.

Hosta 'Golden Meadows' (Van Elderen NR)

[Sport of *Hosta* 'Elegans']

A non registered large size hosta. Leaves are undulated, yellow green with a blue margin. That margin becomes dark green later in the season.

Hosta 'Great Arrival'

M. Zilis (OIR) A. Summers (N) 2003

[Sport of *Hosta* 'Great Expectations']

A large size hosta. Leaves are blue green with a yellow margin. Leaves are deeply corrugated, wavy and moderately cupped. Leaves are glaucous on top and underneath. This hosta looks like *Hosta* 'Great Expectations'. This hosta is more resistant to sun burns.

Hosta 'Great Expectations'

P. Aden (R) 1988

[Sport of *Hosta* 'Elegans']

A large slow growing hosta. Leaves are chartreuse with a dark green margin. As it is a slow grower, water more and fertilize to speed up its maturity. In colder areas, mulch this hosta. The new growth have tendency to freeze.

Hosta 'Northern Exposure'

Walters Gardens, Inc. (ONIR) 1997

[Sport of *Hosta* 'Elegans']

A large hosta. Leaves are medium blue green with a creamy white margin. Leaves are glaucous on top and underneath. This hosta looks like *Hosta* 'American Halo' or *Hosta* 'Northern Halo'.

Hosta 'Northern Halo'

Walters Gardens, Inc. (R) 1984

[Sport of *Hosta* 'Elegans']

A large hosta. Leaves are blue green with a creamy white margin. This hosta looks like *Hosta* 'American Halo' or *Hosta* 'Northern Exposure'.

Hosta 'Piedmont Gold'

F. H. Payne (R) 1982

[Parentage unknown but we are sure it has genes from *Hosta sieboldiana*]

A medium size gold yellow hosta. This hosta doesn't seem to be easily burnt by the sun. This hosta is a good plant for hybridizing. This hosta has to be check for its tendency to produce sports and give very nice new hostas.

Hosta 'Queen of Hearts'

R. Savory (R) 1983

[Sport of a hybrid of *Hosta sieboldiana*]

A medium size hosta. Leaves are medium green with a creamy white margin. This hosta looks like *Hosta* 'Carnival'.

Hosta 'Satisfaction'

C. Wasitis (ON) Bridgewood Gardens (IR) 2000

[Sport of *Hosta* 'Piedmont Gold']

A large hosta. Leaves are dark green with a yellow margin. Leaves are slightly rippled, lightly shiny on top and underneath. Very attractive in a garden.

Hosta 'Thunberbolt'

Hawksridge Farms NR

[Sport du *Hosta* 'Elegans']

A non registered medium size hosta. Leaves are creamy white with a blue green narrow margin. Two hostas look almost the same: *Hosta* 'Dream Queen' and *Hosta* 'Dream Weaver'.

There are other sports of *Hosta sieboldiana* :

- *Hosta* 'Barbara Ann'
- *Hosta* 'Bigga Luigi'
- *Hosta* 'Crème de Menthe'
- *Hosta* 'Doctor Reath'
- *Hosta* 'Foxfire Shadow Dancer'
- *Hosta* 'Forncett Frances'
- *Hosta* 'Golden Meadows'
- *Hosta* 'Hasamona Jeffrey L. Swarts'
- *Hosta* 'Homestead'
- *Hosta* 'Jim Cooper'
- *Hosta* 'Myerscough Magic'
- *Hosta* 'Northern Lights'
- *Hosta* 'Northern Mist'
- *Hosta* 'Northern Star'
- *Hosta* 'Nutty Husband'
- *Hosta* 'Spinning Wheel'
- *Hosta* 'Swiss Lady'
- *Hosta* 'Temptation'

There are many seedlings that show genes of this type resulting from crossings with *Hosta sieboldiana*.

Hosta 'Abiqua Drinking Gourd'

Walden West (R) 1989

[*Hosta* 'Tokudama' × *Hosta sieboldiana*]

A slow growing medium size dark blue hosta. This hosta is quite special as the leaves have a cup shape. After a rainfall you could see birds drinking on the leaves. This hosta looks like *Hosta* 'Silver Bowl' or *Hosta* 'Love Pat'.

Hosta 'Aqua Velva'

K. Vaughn (R) 1983

[A hybrid of a seedling F3 from (*Hosta* 'Polly Bishop' × *Hosta* 'Blue Lace') × *Hosta* 'Summer Fragrance']

A large blue green hosta. This hosta looks like *Hosta* 'Sea Lotus Leaf'.

Hosta 'Big Mama'

P. Aden (R) 1976

[*Hosta* 'Blue Tiers' × *Hosta* 'Blue Angel']

A large slow growing blue hosta. This hosta looks like *Hosta* 'Big Daddy'.

Hosta 'Blue Angel'

P. Aden (R) 1986

[*Hosta* 'Aden 365' × *Hosta* 'Aden 361']

A very large blue hosta. Leaves have several undulations. One of the most giant hosta on the market. This hosta will give a lot of sports.

Hosta 'Blue Betty Lou'

C. Owens (R) 1987

[*Hosta* 'True Blue' × *Hosta* 'Gold Regal']

A large blue green hosta. Leaves have slight undulations. This hosta should be a nice introduction in a garden.

Hosta 'Blue Mammoth'

P. Aden (ONI) P. & J. Ruh (R) 2010

[Hybrid of *Hosta sieboldiana*]

A very large blue green hosta. Leaves are lightly corrugated and lightly folded. Leaves are glaucous on top and dull underneath. Like *Hosta* 'Blue Angel' it should be a nice introduction in a garden.

Hosta 'Blue Umbrellas'

P. Aden (R) 1978

[*Hosta* 'Tokudama' × *Hosta* 'Elegans']

A large blue green hosta. This hosta is very popular in the gardens. This hosta looks like *Hosta* 'Millennium'.

Hosta 'Bressingham Blue'

Alan Bloom (R) 1984

[Hybrid of *Hosta. sieboldiana*]

A large blue green hosta. Leaves have large undulations. This hosta looks like *Hosta* 'Elegans'.

Hosta 'Brigadier'

M. Zilis (ONIR) 2003

[*Hosta* 'Elegans' × *Hosta* 'Hirao Supreme']

A fast growing large medium blue green hosta. Leaves are lightly wavy and corrugated. Leaves are glaucous on top and underneath.

Hosta 'Dorothy Benedict'

R. Benedict (R) 1983

[*Hosta* 'Frances Williams' × *Hosta* 'Frances Williams']

A large blue green hosta. Leaves are streaked yellow chartreuse. This hosta is often used in hybridizing programs

Hosta 'Eleanor J. Reath'

D. Reath (R) 1986

[*Hosta* 'Frances Williams' x *Hosta* 'Elegans']

A large blue green hosta. This hosta looks like *Hosta* 'Blue Mammoth' and *Hosta* 'Ryan's Big One'.

Hosta 'Elegans' aka *Hosta* 'sieboldiana Elegans'

G. Arends (OI) N. Hylander (N) AHS (R) 1987

[Parentage unknown but seem to come from a crossing between *Hosta sieboldiana* and *Hosta* 'Tokudama']

A very large bluish grey green hosta. This hosta is one of the most popular hostas in the gardens. This hosta was used very often in hybridizing. This hosta gives lots of sports. Today, many hostas look alike.

In my garden, this hosta was planted about forty years ago and was never divided.

Hosta 'Freckles'

A Viette (O) A. Summers (NI) AHS, K. Walek (R) 2009

[Parentage unknown but we are sure it has genes from *Hosta sieboldiana*]

A very large hosta. Leaves are green with yellow blotches. Leaves are slightly rippled, lightly corrugated and moderately wavy. This hosta foliage looks like it is virussed.

Hosta 'Gray Cole'

Kuk's Forest Nursery (R) 1985

[A select form of *Hosta. sieboldiana*]

A large blue green hosta. This hosta looks like *Hosta* 'Elegans'.

Hosta 'Halcyon'

E. Smith & BHHS (R) 1988

[Hosta from Tardiana group: *Hosta* 'Tardiflora' x *Hosta sieboldiana*]

A medium size blue hosta. This hosta will be one of the last to go dormant late in the season. Two hostas are look alikes: *Hosta* 'Blue Arrow' and *Hosta* 'Canadian Blue'.

Hosta 'Minnie Klopping'

A. Arett (R) 1975

[Parentage unknown but we are sure it has genes from *Hosta sieboldiana*]

A medium size dark green hosta. Leaves are satiny on top. This hosta will often take a vase shape form.

Hosta 'Mira'

F. Maekawa (O) F. Maekawa & W. G. Schmid (N) Japan (I) AHS (R) 2002

[Parentage unknown but we are sure it has the genes from *Hosta sieboldiana*]

A medium size medium green hosta. Leaves are slightly rippled, lightly cupped, lightly corrugated and lightly wavy. Leaves are slightly shiny on top and dull underneath. Its foliage is greener than *Hosta* 'Elegans'. If the weather permits, it takes nice fall colors.

Hosta 'Montreal Blue Ice Storm'

V. Wade (ONIR) 2007

[Seedling of *Hosta* 'Elegans' × seedling of *Hosta* 'Elegans']

A very large blue green hosta. Leaves are slightly rippled and lightly wavy. Leaves are glaucous on top and underneath. We discovered this hosta in the gardens of Wade and Gatton in Ohio. Hidden between hundreds of hostas, it was never found. Being the firsts to discover it, Van Wade asked us to find a name for it. Remembering the famous ice storm in Montreal area years ago, the name was then given. This hosta was also the selected hosta to celebrate the 10th anniversary of Quebec Hosta and Hemerocallis Society.

Hosta 'President's Choice'

S. Moldovan (R) 1983

[Hybrid of *Hosta sieboldiana*]

A large blue green hosta. Leaves are ruffled.

Hosta 'Quilted Hearts'

M. Soules (R) 1989

[Hybrid of *Hosta sieboldiana*]

A medium size green heavily corrugated hosta. Leaves become very shiny as the season progress.

Hosta 'Robert Frost'

W. & E. Lachman (R) 1988

[*Hosta* 'Banana Sundae' × *Hosta* 'Frances Williams']

A medium size hosta. Leaves are dark blue green with a creamy yellow to near white margin. *Hosta* 'Heart's Content, *Hosta* 'Northern Halo' and *Hosta* 'Showboat' are look alikes.

Hosta 'Rosedale Tractor Seat'

J. Hadrava (ONIR) 1999

[*Hosta* 'Galaxy' × *Hosta* 'Elegans']

A large medium green hosta. Leaves are glaucous on top and underneath. Size and form of the leaves are really like a tractor seat.

Hosta 'Ryan's Big One'

L. Englerth (R) 1982

[Parentage unknown but we are sure it is a seedling from *Hosta* 'Elegans']

A large blue green hosta with some corrugation and slight undulations along the leaf. It looks like *Hosta* 'Elegans'.

Hosta 'Samual Blue'

G. Krossa & P. Ruh (R) 1981

[Parentage unknown but we are sure in come from a crossing between *Hosta sieboldiana* x *Hosta* 'Tokudama']

A medium size powdery blue hosta. It seems that Gus Krossa and Peter Ruh brought this hosta from a trip to Japan in the 1960s.

Hosta 'Sea Blue Monster'

M. Seaver (ONI) AHS, K. Walek (R) 2009

[*Hosta* 'Brookwood Blue' x an unknown hosta]

A large blue green hosta. Leaves are moderately corrugated and convexly cupped. Another hosta that looks like *Hosta* 'Elegans'.

Hosta 'Sea Lotus Leaf'

M. Seaver (R) 1985

[Hybrid of *Hosta* 'Wagon Wheels']

A large blue green hosta. Leaves are cupped with good substance. Slug resistant.

Hosta 'Sea Monster'

M. Seaver (R) 1978

[Hybrid of *Hosta* 'Brookwood Blue']

A large bright green hosta. One of the rare green hostas from the type *sieboldiana*.

Hosta 'Silver Bowl'

E. Fisher & AHS (R) 1986

[Parentage unknown but we are sure it is a seedling of *Hosta sieboldiana*]

A large blue green hosta. Hosta offering a nice combination with the color of the foliage and its cupped shape.

Hosta 'Sunlight Sister'

E. Nesmith & AHS (R) 1986

[Parentage unknown but we are sure it is a seedling of *Hosta sieboldiana*]

A large yellow green to chartreuse hosta. It doesn't seem to suffer from sun burns in spring. This hosta looks like *Hosta* 'Piedmont Gold'.

Hosta 'Willy Nilly'

G. Holly (ONI) P. Ruh (R) 1996

[Parentage unknown but we are sure it is a seedling of *Hosta* 'Elegans']

A large gray green hosta. Leaves are slightly rippled. Leaves are waxy on top and dull underneath. It could be very large at maturity.

Hosta 'Zounds'

P. Aden (R) 1978

[*Hosta* 'Golden Waffles' × *Hosta* 'Golden Prayers']

A medium size chartreuse gold hosta. This hosta doesn't seem to suffer from sun burns in spring. This hosta stays nice all season.

There are many other hostas seedlings having *hosta sieboldiana* as parent

- *Hosta* 'Abiqua Pagoda'
- *Hosta* 'Abiqua Parasol'
- *Hosta* 'Agawa'
- *Hosta* 'Alabama Bowl'
- *Hosta* 'American King of the Woods'
- *Hosta* 'American Super Elegans'
- *Hosta* 'Birchwood Blue'
- *Hosta* 'Bix Blues'
- *Hosta* 'Blue Mountains'
- *Hosta* 'Blue Plisse'
- *Hosta* 'Blue Rock'
- *Hosta* 'Blue Saucers'
- *Hosta* 'Blue Seer'
- *Hosta* 'Blue Thunder'
- *Hosta* 'Bold Ruffles'
- *Hosta* 'Brookwood Blue'
- *Hosta* 'Buckshaw Blue'
- *Hosta* 'Clarence'
- *Hosta* 'Crater Lake'
- *Hosta* 'Dorset Charm'
- *Hosta* 'Dorset Flair'
- *Hosta* 'Edina Heritage'
- *Hosta* 'Ellen F. Weissenberger'
- *Hosta* 'Filagree'
- *Hosta* 'Fond Hope'
- *Hosta* 'Glacial Lake'
- *Hosta* 'Gold Seer'
- *Hosta* 'Golden Nugget'
- *Hosta* 'Green Saucer'
- *Hosta* 'Happiness'
- *Hosta* 'Helen Doriot'
- *Hosta* 'Herb Altman'
- *Hosta* 'Honey Hill Water Ladle'
- *Hosta* 'Housatonic'
- *Hosta* 'Irische See'
- *Hosta* 'Khaki Krush'
- *Hosta* 'Kiwi Blue Ruffles'
- *Hosta* 'Lady in Waiting'
- *Hosta* 'Lee Armiger'
- *Hosta* 'Lilac Giant'
- *Hosta* 'Metallic Sheen'
- *Hosta* 'Mike Shadrack'
- *Hosta* 'Mira'
- *Hosta* 'Nesmith Giant'
- *Hosta* 'Nightlife'
- *Hosta* 'Ocean Tide'
- *Hosta* 'Paradise Blue Sky'
- *Hosta* 'Perry's True Blue'
- *Hosta* 'Prince of Wales'
- *Hosta* 'Puckered Giant'
- *Hosta* 'Regal Ruffles'
- *Hosta* 'Quilted Skies'
- *Hosta* 'Rembrandt Blue'
- *Hosta* 'Ritz'
- *Hosta* 'Roscoe'
- *Hosta* 'Rosedale Big Bopper'
- *Hosta* 'Rosedale Black Ice'
- *Hosta* 'Rosedale Dewy'
- *Hosta* 'Rosedale Pebbles'

- *Hosta* 'Rosedale Pickles'
- *Hosta* 'Rosedale Terraced Hills'
- *Hosta* 'Rosedale Yellow Brick Road'
- *Hosta* 'Roy Lee Goodenough'
- *Hosta* 'Sea Blue'
- *Hosta* 'Silver Award'
- *Hosta* 'Simply Sharon'
- *Hosta* 'Skookumchuck'
- *Hosta* 'Sky Kissed'
- *Hosta* 'Snowden'
- *Hosta* 'Standing Ovation'
- *Hosta* 'Stonewall'
- *Hosta* 'Summer Snow'
- *Hosta* 'Sumi'
- *Hosta* 'Sunrise Crinkles'
- *Hosta* 'Waterford'
- *Hosta* 'Woodland Blue'
- *Hosta* 'Wrinkles and Crinkles'

To summarize, *Hosta sieboldiana* and all the ones from that type are not afraid of long winters with a lot of snow. They must be planted in part shade areas. The majority of these hostas will be quite large.

Hosta 'Tokudama' type

Hosta 'Tokudama' was originaly considered as species of the gender Hosta.

It is in 1991 that this hosta became a cultivar related to *Hosta sieboldiana*.

Hostas related to *Hosta* 'Tokudama' came from a series of crossings between inter-species of *Hosta sieboldiana*. They must be planted in shady areas, because they don't have any resistance to the sun. Leaves have a cupped shape form and have good substance and this thickness makes them slug resistant.

One of the weak points of these hostas is their growing speed which is very slow, but they are very strong and hardy. They must be fertilized and given a good humid soil to get best results. However the hostas from this type will all be small.

This hosta is often crossed with large hostas to get all kinds of new sizes.

Plants being sports of *Hosta* 'Tokudama'.

Hosta 'Blue Shadows'

K. Anderson (R) 1980

[A select form of *Hosta* 'Tokudama Aureonebulosa']

A medium size blue green hosta. Leaves are streaked with yellow chartreuse,

Hosta 'Golden Bullion'

P. Bennerup & P. Ruh (R) 1989

[Sport of *Hosta* 'Tokudama Flavocircinalis']

A medium size yellow green hosta. This hosta is very slug resistant like other hostas from that group.

Hosta 'Golden Medallion'

M. Eisel & AHS (R) 1984

[Sport of *Hosta* 'Tokudama Aureonebulosa']

A medium size yellow green hosta. This hosta is very slug resistant like other hostas from that group.

Hosta 'Tokudama Aureonebulosa'

F. Maekawa & AHS (R) 1987

[Sport of *Hosta* 'Tokudama']

A medium size hosta. Leaves are greenish yellow with a blue green margin. A slug resistant hosta. Even if it has a slow growing speed, this hosta is very useful to hybridizers.

Hosta 'Tokudama Flavocircinalis'

F. Maekawa & AHS (R) 1987

[Sport of *Hosta* 'Tokudama']

A medium size hosta. Leaves are blue green with a yellow margin. The colors of this hosta are the reverse of *Hosta* 'Tokudama Aureonebulosa'. This hosta has a faster growth speed than the other hostas of that type.

Hosta 'Tokudama Flavoplanata'

F. Maekawa & AHS (R) 1987

[Sport of *Hosta* 'Tokudama']

A medium size hosta. Leaves are medium yellow with a blue green margin. The margin is so narrow that we have difficulty to see it.

Among the other sports of *Hosta* 'Tokudama' there are *Hosta* 'Kiwi Forest' and *Hosta* 'Smokey Tokey'.

Today we have many other plants coming from crosses with

Hosta 'Tokudama'

Hosta 'Aspen Gold'

V. Grapes & AHS (R) 1986

[Parentage unknown but we are sure that it has genes from *Hosta* 'Tokudama']

A medium size chartreuse yellow hosta. This hosta is very slow to reach maturity.

Hosta 'Betcher's Blue'

AHS (R) 1986

[Hybrid of *Hosta* 'Tokudama']

A medium size blue green hosta. This hosta is remarkable by the large undulations on the margin of the leaves that are cupped.

Hosta 'Lime Krinkles'

M. Soules (R)

[Hybrid of *Hosta* 'Tokudama']

A medium size chartreuse hosta. Leaves are remarkable with the large undulations.

Hosta 'Love Pat'

P. Aden (R) 1978

[*Hosta* 'Blue Velvet' × *Hosta* 'Blue Vision']

A medium size intensely blue hosta. This hosta is a slow grower. This hosta looks like *Hosta* 'Abiqua Drinking Gourd'.

Hosta 'Moscow Blue'

A Arett & AHS (R) 1986

[Parentage unknown but we are sure it has genes from *Hosta* 'Tokudama']

A medium size blue green hosta. This hosta will keep its blue color quite late in the season.

Hosta 'Prairie Fire'

Kuk's Forest Nursery (OR) 1994

[*Hosta* 'Gold Regal' × a yellow seedling of *Hosta* 'Tokudama']

A medium size deep gold hosta. Leaves are slightly rippled, dull on top and glaucous underneath. It will take many years to be mature.

Hosta 'Quilted Cup'

E. Minks (R) 1980

[*Hosta* 'Tokudama' × *Hosta* 'Crinkle Cup']

A medium size blue green hosta. Many of the leaves will take a cupped form, while others will be flat.

Hosta 'Serendipity'

P. Aden (R) 1978

[*Hosta* 'Blue Cadet' × *Hosta* 'Tokudama']

A medium size blue green hosta. Due to its fast growing speed, it can be used as ground cover.

Hosta 'Van Wade Blue'

M. Zilis (ONR) V. Wade (I) 2003

[Parentage unknown but we are sure that it has genes from *Hosta* 'Tokudama']

A medium size intensely blue green hosta. Leaves are deeply corrugated, moderately cupped and lightly wavy. This hosta looks like *Hosta* 'Blue Vision.

Hosta 'Wagon Wheels'

E. Minks (R) 1971

[Sport of *Hosta* 'Golden Nugget']

A medium size hosta. Leaves are gray green with a chartreuse margin. This hosta looks like *Hosta* 'Frances Williams' but smaller. It will show the burn signs too.

Among the seedlings there are

- *Hosta* 'Bengee'
- *Hosta* 'Blue Dome'
- *Hosta* 'Blue Horizon'
- *Hosta* 'Blue Lagoon'
- *Hosta* 'Blue Troll'
- *Hosta* 'Bon Voyage'
- *Hosta* 'Carder Blue'
- *Hosta* 'Colesburg Blue'
- *Hosta* 'Estelle Aden'
- *Hosta* 'Evelyn McCafferty'
- *Hosta* 'Fleeta's Blue'
- *Hosta* 'Fleeta's Brownell Woodroffe'
- *Hosta* 'Golden Chimes'
- *Hosta* 'Golden Fleece'
- *Hosta* 'Golden Rajah'
- *Hosta* 'Hallie's Honor'
- *Hosta* 'Hirao Zeus'
- *Hosta* 'Johanne'
- *Hosta* 'Julia Hardy'
- *Hosta* 'Kiwi Dreadlocks'
- *Hosta* 'Krinkled Joy'
- *Hosta* 'Lakeside Sapphire Pleats'
- *Hosta* 'Like Butta',
- *Hosta* 'Mayan Seer',
- *Hosta* 'Mood Indigo',
- *Hosta* 'Mogul',
- *Hosta* 'Mrs. Toshie',
- *Hosta* 'Northwest Textures'
- *Hosta* 'Patricia's Paragon',
- *Hosta* 'Pelham Blue Tump',
- *Hosta* 'Pistachio Glow',
- *Hosta* 'Rod's Gold',
- *Hosta* 'Rough Waters',
- *Hosta* 'Sea Gold Dust',
- *Hosta* 'Sea Grotto',
- *Hosta* 'Seaside Splendor',
- *Hosta* 'Seersucker',
- *Hosta* 'Silver Rim',
- *Hosta* 'St. George's Gold',
- *Hosta* 'Tisch',
- *Hosta* 'Treasure',
- *Hosta* 'Ur',
- *Hosta* 'Warwick Cup',
- *Hosta* 'Wayside Blue'
- *Hosta* 'Zager's Tokudama'.

To summarize: we know that the hostas from the crosses resulting in Hosta 'Tokudama' have leaves that are round and cupped shaped. They are slow growers but the plants are very strong. These hostas are hardy and slug resistant. They must be planted in shady areas.

Hosta nigrescens species

Hosta nigrescens

This species comes from the northern part of Japan. It doesn't seem that we discovered this plant in the wild, but similar plants were present in private gardens in that region. This hosta species is different by its high et wide vase shape form. The foliage has a good substance is corrugated but there is only one leaf per division. The flowers are pale lavender.

Identified by Makino in 1902

A large vase shaped green hosta.This species doesn't produce many sports and if it does, sports have often yellow or variegated foliage.

The only sport known so far is *Hosta 'Victory'*.

Hosta 'Victory'

M. Zilis (OR) M. Zilis & R. Solberg (NI) 2003

[Sport of *Hosta nigrescens* 'Elatior']

A large vase shaped hosta. Leaves are medium green with a creamy yellow margin. Leaves are flat, very shiny on top and underneath. This hosta looks like *Hosta* 'Sagae'.

Today we have many seedlings coming from crossings with *Hosta nigrescens*.

Hosta 'Elatior'

F. Maekawa (ONI) P. & J. Ruh (R) 2002

[Parentage unknown but we are sure it has the genes from *Hosta nigrescens*]

A large vase shape green hosta. Leaves are shiny. This hosta can be a focal point in a garden.

Hosta 'Flower Power'

K. Vaughn (R) 1987

[*Hosta nigrescens* × *Hosta plantaginea*]

A large gray green hosta. Leaves are slightly ruffled. This hosta shows the different traits of both parents.

Hosta 'Garnet Prince'

P. Banyai & J. Dishon (R) 1991

[Hybrid of *Hosta nigrescens*]

A large green hosta. Leaves have satiny sheen on top. The petiole is also bright cherry red.

Hosta 'Key West'

O. Petryszyn (ONIR) 1999

[*Hosta nigrescens* 'Elatior' × *Hosta* 'High Noon']

A large greenish yellow hosta. Leaves are slightly rippled. Leaves are slightly shiny on top and dull underneath. This hosta can be the star in a garden.

Hosta 'Krossa Regal'

G. Krossa & A. Krossa (R) 1980

[Hybrid of *Hosta nigrescens*]

A large vase shapes blue gray hosta. Leaves have large undulations. Superb plant in any landscaping. *Hosta* 'Regal Splendor' is a sport of this nice hosta.

Hosta 'Spotlight'

J. Wilkins (ONI) J. & J. Wilkins (R) 1995

[*Hosta* 'Fort Knox' × *Hosta nigrescens* 'Elatior']

A large medium gold hosta. Leaves are smooth, very shiny on top and underneath.

Hosta 'True Blue'

P. Aden (R) 1978

[*Hosta* 'Chartreuse Wedge' × (*Hosta nigrescens* x *Hosta* 'Blue Vision')]

A large blue hosta. Even if it is in the *Hosta nigrescens* family, this hosta looks like *Hosta sieboldiana*.

Other seedlings are

- *Hosta* 'All That Jazz',
- *Hosta* 'Allegan Emperor',
- *Hosta* 'Green Wedge',
- *Hosta* 'John Wargo'
- *Hosta* 'Nate the Great'
- *Hosta* 'Tenryu'

If we summarize, these hostas are slug resistant due to their thick leaves. They also resist to wind. They will look great all season long.

Hosta fluctuans species

Hosta fluctuans, when discovered, it was growing in a habitat located in the hilly forests along the west coast of Japan, where the mountains reach the Pacific Ocean. In that region, the rain season starts in July and many elevated parts are covered with snow from December till April. We don't see that species in our gardens.

In the open areas, this species is often competing with tall grasses or small shrubs. But it always survives with them.

For many years, the leaves of this species were used in floral arrangements (Ikebana).

This hosta species is different by its erect form, tall and wide. The leaves are wide and thick, slightly undulated, shiny on top. Very slug and wind resistant. These hostas will be nice until the first frosts. Flowers are pale lavender.

Plants being sports of *Hosta fluctuans*.

Hosta 'Sagae'

K. Watanabe (N) AHS (R) 1996

[Sport of *Hosta fluctuans*]

A large erected hosta. Leaves are yellow green with a pale yellow margin. Leaves have irregular undulations. This hosta might give many sports. This hosta is used regularly by hybridizers to create hostas with thick leaves. Excellent plant in a garden.

Hosta 'Clifford's Forest Fire'

J. Clifford (ONIR) 2000

[Sport of *Hosta* 'Sagae']

A large erected hosta. Leaves are slightly blue green with a yellow margin. Leaves are slightly rippled and slightly shiny on top. This hosta looks like *Hosta* 'Magic Fire'. It is also similar to *Hosta* 'Sagae' except for the margin that is larger than *Hosta* 'Sagae'.

Hosta 'Liberty'

J. Machen, Jr. (ONIR) 2000

[Sport of *Hosta* 'Sagae']

A large erected hosta. Leaves are medium blue green with a yellow margin. As soon this hosta reached the nurseries it became very popular.

Hosta 'Majesty'

R. Garbe (ONIR) 1999

[Sport of *H.* 'Sagae']

A large erected hosta. Leaves are medium green with a pale yellow margin. Leaves are slightly rippled. This hosta looks like *Hosta* 'Liberty'.

Among the other sports, there are *Hosta* 'Emerald Emperor', *Hosta* 'Ginryu', *Hosta* 'Ivory Coast' and *Hosta* 'Magic Fire'

Today we have many seedlings coming from crosses with *Hosta fluctuans*.

Hosta 'Glacial Lake'

Walden West (OR) J. Hyslop (NI) 1999

[*Hosta* 'Sagae' × *Hosta* 'Elegans']

A large erected medium blue green hosta. Leaves are slightly rippled. Leaves are glaucous on top and underneath.

Hosta 'Jubilation'

J. Wilkins (OIR) J. & S. Wilkins (N) 2000

[*Hosta* 'HB-87-1' × *Hosta* 'Sagae']

A large erected hosta. Leaves are dark green with a greenish yellow margin. Leaves are slightly rippled, slightly shiny on top and glaucous underneath. This hosta was seen in the hybridizer's garden.

Hosta 'Kiwi Skyscraper'

Barry Slight NR

[*A seedling from Hosta* 'Sagae']

A non-registered hosta from Barry Sligh in New Zealand. A very large erected blue green hosta.

Hosta 'Spacious Skies'

Walden West (OIR) J. Hyslop (N) 1999

[Hybrid of *Hosta* 'Sagae']

A large erected intensely blue green. The leaves are slightly rippled. Leaves are glaucous on top and underneath.

Among the other seedlings there are

- *Hosta* 'Big Zach'
- *Hosta* 'Dark Tower'
- *Hosta* 'Fat Cat'
- *Hosta* 'Gunsmoke'
- *Hosta* 'H.D. Thoreau'
- *Hosta* 'Ivory Tower'
- *Hosta* 'Lakeside Surf Rider'
- *Hosta* 'Margaret Eyre'
- *Hosta* 'Mesa Verde'
- *Hosta* 'Midnight Express'
- *Hosta* 'Thank You'
- *Hosta* 'Vanguard'
- *Hosta* 'Yojimbo'

To summarize: these hostas have large undulated and thick leaves. They are wind and slug resistant. These plants will be nice till first frosts.

Hosta ventricosa species

This species was living in the same conditions as the *Hosta plantaginea,* as it tolerates more heat and humidity. This hosta requests a sunny area and lots of water in August. Due to its tolerance to heat and humidity, that favors the growth of new leaves during the season.

The hostas of the type *ventricosa* emerge quite early in spring. So there is a possibility that in the cold regions, risks of late frosts are always possible. Protecting them with mulch is a good habit.

This species is less popular to the hybridizers. As this species is the only tetraploid hosta, we can use it as a mother plant but it will be better to give its pollen, being excellent to favor streaked purple flowers.

This hosta could grow quite large when mature. The leaves are dark green in a heart shape.

Plants being sports of *Hosta ventricosa*.

Hosta ventricosa 'Aureomarginata'

K. Hensen & AHS (R) 1986

[Sport of *Hosta ventricosa*]

A medium size hosta. Leaves are dark emerald green with a yellow margin.

Some other sports are *Hosta ventricosa* 'Aureomaculata', *Hosta* 'Crystal Fountain', *Hosta* 'Gold Flush', *Hosta* 'Ogon Murasaki' and *Hosta* 'Tiffney's Dart Vader'.

But there are many seedlings coming from crosses with

Hosta ventricosa:

Hosta 'Beauty Little Blue'

H.J. van den Top NR

[*Hosta* Blue Cadet X Hosta clausa normalis]

A small non-registered blue green hosta. This hosta displays the best characteristics of both parents.

Hosta 'Betty'

R. Benedict (R) 1983

[*Hosta nakaiana* × *Hosta ventricosa*]

A small dark green hosta. Leaves are ruffled. This hosta looks like *Hosta ventricosa* but smaller.

Hosta 'Heartache'

R. Benedict (ONI) AHS, K. Walek (R) 2009

[*Hosta* 'Gold Regal' × *Hosta ventricosa*]

A small golden yellow hosta. Leaves are lightly corrugated, dull on top and glaucous underneath.

Hosta 'Kiwi Black Magic'

B. Sligh (ONIR) 1999

[*Hosta sieboldiana* × *Hosta ventricosa*]

A small dark green hosta. Leaves are slightly rippled, dull on top and slightly shiny underneath. This hosta looks like *Hosta* 'Lakeside Black Satin'.

Hosta 'Lakeside Black Satin'

M. Chastain (OR) 1993

[Hybrid of *Hosta ventricosa*]

A small very dark green hosta. Leaves are slightly rippled, slightly shiny on top and dull underneath. To me it is the darkest hosta seen in a garden.

Hosta 'Lakeside Ebony Echoes'

M. Chastain (ONIR) 2002

[*Hosta ventricosa* × *Hosta ventricosa*]

A medium size dark green hosta. Leaves are rolled under, lightly cupped down and twisted at the tip.

Hosta 'Montreal'

Le Jardin Botanique de Montreal (R) 1983

[Parentage unknown but we are sure it has the genes of *Hosta ventricosa*]

A medium size intensely blue green hosta. Leaves are slightly rippled and lightly wavy. This hosta is a fast grower. One of the three hostas registered by Montreal botanical Garden.

Hosta 'Mount Royal'

Le Jardin Botanique de Montreal (R) 1983

[Parentage unknown but we are sure it has the genes of *Hosta ventricosa*]

A medium size intensely pale green hosta. This hosta is a fast grower. One of the three hostas registered by Montreal botanical Garden.

Hosta 'Slick Willie'

J. Hadrava (OR) 1996

[*Hosta ventricosa* × *Hosta* 'Invincible']

A medium size dark green hosta. Leaves are smooth, very shiny on top and underneath. This hosta looks like *Hosta* 'Lakeside Black Satin'.

Hosta 'Sunny Disposition'

F. Nyikos (ONIR) 2001

[*Hosta ventricosa* × *Hosta* 'Birchwood Parky's Gold']

A small greenish yellow hosta. Leaves are moderately heavily rippled, lightly twisted, folded and wavy. The foliage might burn under full sun.

Hosta 'Waving Wuffles'

T. Avent (OR) 1995

[Hybrid of *Hosta ventricosa*]

A small medium green hosta. Leaves are heavily rippled. Leaves are slightly shiny on top and underneath.

Other seedlings are

- *Hosta* 'First Knight'
- *Hosta* 'Holly's Honey'
- *Hosta* 'Honey Hill Pins and Needles'
- *Hosta* 'Jolly Green Dwarf'
- *Hosta* 'John Metzgar'
- *Hosta* 'Lakeside Emerald Lights'
- *Hosta* 'Mama's Gold'
- *Hosta* 'Matthew J. Walton'
- *Hosta* 'Odyssey'
- *Hosta* 'PeeDee Elfin Bells'
- *Hosta* 'PeeDee Picotee'
- *Hosta* 'PeeDee Treasure'
- *Hosta* 'Rachel'
- *Hosta* 'Rosedale Barnie'
- *Hosta* 'Rosedale Dough Boy'
- *Hosta* 'Rosedale Monica Havana'
- *Hosta* 'Rosedale Mrs. President'
- *Hosta* 'Rosedale Stiched in Silk'
- *Hosta* 'Superba'
- *Hosta* 'Taiwan'
- *Hosta* 'Tall Boy'
- *Hosta* 'Tardiflora Hybrida'
- *Hosta* 'Tucker Valentine'
- *Hosta* 'Twist of Fate'
- *Hosta* 'Unducosa'
- *Hosta* 'Unique'
- *Hosta* 'Venucosa'

If we summarize, if you want to have success with hosta of this type Hosta ventricosa, plant them in a sunny area with lots of water in August. These hostas will produce new leaves all season long. Mulching is a good habit to protect them.

Hosta hypoleuca species

This species was found in the Province of Mikawa in Japan. It is growing in full sun areas, on the south face of a canyon. It was attached to the side of the cliffs. It was also seen on vertical walls of ravines in deep valleys. The white coating on the back of the leaves is an adaptation to protect them against the extreme heat caused by the reflection on the rocky surface. This hosta can grow easily to 4 feet (1m20) in diameter.

Hosta hypoleuca will prefer a sunny area and can tolerate temperatures as low as 45° F (7° C). It likes sandy and/or clay soil, but with good drainage. It is a good idea to protect it with mulch in winter. To show the white coating under the leaves it should be planted in a high location in the garden.

Flowers of this species are very pale lavender. This plant is very appreciated by hybridizers. This hosta will also produce many sports.

In Japan, *Hosta pycnophylla* was found in the Setouchi Inland sea region. This hosta is very similar to *Hosta hypoleuca*, it also has the white coating under the leaves.

Hosta hypoleuca will not produce many sports.

Many other seedlings had the genes from *Hosta hypoleuca*.

Hosta 'A Many-splendored Thing'

E. Elslager (ONIR) 1999

[*Hosta* 'Dorothy Benedict' × *Hosta hypoleuca*]

A medium size hosta. Leaves are dark green with a golden yellow margin. Leaves are slightly shiny on top and glaucous underneath. This hosta looks like *Hosta* 'Band of Gold'.

Hosta 'Arc de Triomphe'

E. Elslager (ONIR) 2000

[*Hosta* 'Foxy Doxy' × *Hosta hypoleuca*]

A medium size hosta. Leaves are dark green with a yellow margin. Leaves are slightly rippled. Leaves are slightly shiny on top and glaucous underneath. Often margin becomes white during summer.

Hosta 'Arch Duke'

R. Lydell (OIR) 2000

[*Hosta hypoleuca* × *Hosta* 'Cartwheels']

A medium size light to medium green hosta. Leaves are slightly rippled, dull on top and glaucous underneath.

Hosta 'Azure Snow'

P. Ruh (R) 1991

[Hybrid of *Hosta hypoleuca*]

A medium size blue green hosta. Leaves have several undulations. This fast growing hosta is appreciated with its blue color.

Hosta 'Band of Gold'

E. Elslager (ONIR) 1999

Hosta Band of Gold 13314.JPG

[*Hosta* 'Dorothy Benedict' × *Hosta hypoleuca*]

A medium size hosta. Leaves are dark green with a yellow margin. Leaves are slightly rippled, dull on top and glaucous underneath. Petiole of this hosta is yellow marginated. This hosta looks like *Hosta* 'A Many Splendored Thing'.

Hosta 'Butternut Hill'

L. Englerth (ONI) P. & J. Ruh (R) 2001

[*Hosta hypoleuca* × *Hosta hypoleuca*]

A medium size yellow green hosta. Leaves are slightly rippled and moderately wavy. Leaves are shiny on top and dull underneath. During the season, this hosta is getting darker to end the season completely green.

Hosta 'Celestial'

E. Elslager (ONIR) 2000

[*Hosta* 'Dorothy Benedict' × *Hosta hypoleuca*]

A medium size erected hosta. Leaves are dark green with a greenish yellow margin. Leaves are slightly rippled, shiny on top and glaucous underneath.

Hosta 'Kiwi Hippo'

L. Phillips (ON) B. Sligh (IR) 1999

[*Hosta hypoleuca* × *Hosta longipes*]

A medium size slightly blue green. Leaves are slightly rippled, dull on top and glaucous underneath.

Hosta 'Lemon Meringue'

P. Ruh (R) 1991

[Hybrid of *Hosta hypoleuca*]

A small light yellow hosta. Leaves have small undulations. This hosta looks like *Hosta hypoleuca* except for the color.

Hosta 'Merry Sunshine'

E. Elslager (ONIR) 1998

[*Hosta* 'Fascination' × *Hosta hypoleuca*]

A small hosta. Leaves are medium green with a creamy white margin. Leaves are slightly rippled, slightly shiny on top and waxy underneath. Petiole is creamy white edged with streaks.

Hosta 'Ooh La La'

E. Elslager (ONIR) 2000

[*Hosta* 'Sea Prize' × *Hosta hypoleuca*]

A small very hardy hosta. Leaves are creamy white with a dark green margin. Leaves are slightly rippled, slightly shiny on top and glaucous underneath. Petiole is streaked.

Many other seedlings have genes from *Hosta hypoleuca*.

- *Hosta* 'A Blue Streak'
- *Hosta* 'A Lady in Blue'
- *Hosta* 'All Star'
- *Hosta* 'Aqua Duchess'
- *Hosta* 'Big Top'
- *Hosta* 'Blue Ridge Mountains'
- *Hosta* 'Blueberry Pancakes'
- *Hosta* 'Bluegrass Country'
- *Hosta* 'Cascading Waters'
- *Hosta* 'Clambake'
- *Hosta* 'Color a la Mode'
- *Hosta* 'Fickle Blue Genes'
- *Hosta* 'Final Answer'
- *Hosta* 'Floppy Disk'
- *Hosta* 'Ganny Sandy'
- *Hosta* 'Gazpacho'
- *Hosta* 'Glacier Cascade'
- *Hosta* 'H.D. Thoreau'
- *Hosta* 'Happy Valley'
- *Hosta* 'High-rise'
- *Hosta* 'Honey Hill Wrinkled Rainslicker'
- *Hosta* 'Hottie' 'King David'
- *Hosta* 'Koyurugi'
- *Hosta* 'Lakeside Breaking News'
- *Hosta* 'Lasting Impression'
- *Hosta* 'Lighthouse'
- *Hosta* 'Limestone Lover'
- *Hosta* 'Maggie May'
- *Hosta* 'Magnificent Obsession'
- *Hosta* 'Old Black Magic'
- *Hosta* 'Olive Oil'
- *Hosta* 'Roller Coaster'
- *Hosta* 'Sails at Sunset'
- *Hosta* 'Skyward'
- *Hosta* 'Splendor'
- *Hosta* 'Stellar by Starlight'

To summarize: *Hosta hypoleuca* and the hostas from that type will grow well in sunny areas. They can survive drought periods and flooding. These hostas can grow attached on cliffs.

Hosta venusta species

Hosta venusta was transferred from south west Korea to Cheju Island after the last glaciation about 13-14,000 years ago. At the time the ices melted from that volcanic island.

Hosta venusta is a miniature hosta forming a very small green plant. It is the smallest species. Leaves are oval, undulated and shiny. Being quite low, 5" (12 cm) in height, it can reach 9" (20 cm) in diameter. It multiplies very rapidly. With time this plant can spread widely. The dense foliage will stop the weeds.

Hosta venusta is a shade plant and should not be planted full sun. This is a good plant for shady rock gardens. The foliage is not slug resistant. Flowers are lavender. This hosta can be propagated by divisions. This species is very important for hybridizers targetting miniature hostas.

Plants being sports of *Hosta venusta*.

Hosta 'Kinbotan'

Japan (ONI) P. Ruh (R) 2002

[Sport of *Hosta venusta*]

A very small hosta. Leaves are light green with a yellow margin. Leaves are lightly cupped, dull on top and slightly shiny underneath. Petiole is short with faint yellow edge. This hosta can be used in troughs, containers and in rock gardens in shady areas.

Hosta 'Pinwheel'

R. Savory (R) 1983

[Sport of *Hosta venusta*]

A small hosta. Leaves are variagated white with a dark green margin. Leaves will be twisted when mature. This hosta looks like *Hosta* 'Masquerade'.

Hosta 'Striker'

R. Lydell (ONIR) 2000

[Sport of *Hosta venusta*]

A small hosta. Leaves are medium green with a white margin. Leaves are slightly rippled, slightly shiny on top and dull underneath. This hosta looks like *Hosta* 'Little White Line'.

There are other sports from that group: *Hosta* 'Crater's Heart', *Hosta* 'Crater's Rim', *Hosta* 'Shironakafu Otome' and *Hosta* 'Tsugaru Otome'.

There are some seedlings from crossings with genes from *Hosta venusta*.

Hosta 'Abiqua Miniature'

Walden West (R) 1988

[Parentage unknown but we are sure it has the genes from *Hosta venusta*]

A small emerald green hosta. This hosta looks like the species.

Hosta 'Awesome'

H. & D. Benedict (ONI) R. Kuenster, AHS (R) 2010

[A seedling from *Hosta venusta*]

A small slow growing medium green hosta. Leaves are dull on top and underneath. Petiole has some red specklings. This hosta is an excellent choice for troughs and rock gardens in shady areas.

Hosta 'Blue Eyes'

R. Herman (ONIR) 1996

[Hybrid of *Hosta venusta*]

A small medium blue green hosta. Leaves are glaucous on top and underneath. This hosta looks like *Hosta* 'Blue Ice'.

Hosta 'Cody'

R. Solberg (ONIR) 1996

[*Hosta* 'Shining Tot' × *Hosta venusta*]

A small dark green hosta. Leaves are very shiny on top and underneath. This hosta looks like *Hosta* 'Raspberry Sorbet' and *Hosta* 'Tot Tot'.

Hosta 'Craig's Temptation'

G. Shenk & J. Craig (O) G. Schenk (N) G. Shenk (I) P. Ruh (R) 2002

[Probably *Hosta venusta* x *Hosta nakaiana*]

A small slow growing medium green hosta. Leaves are waxy on top, slightly shiny underneath and moderately wavy. Petiole is long.

Hosta 'Elsley Runner'

Inconnu (ONI) AHS, K. Walek (R) 2009

[Seedling from *Hosta venusta*]

A small slow growing green hosta. Leaves are flat and dull on top.

Hosta 'Gold Drop'

K. Anderson (R) 1977

[Hybrid of *Hosta venusta*]

A small golden hosta. This small hosta is often used in front of flower beds or as ground cover.

Hosta 'Hush Puppie'

T. Avent (ONIR) 2005

[(*Hosta* 'Beatrice' x seedling) x (Seedling × *Hosta venusta*)]

A small fast growing hosta. Leaves are medium green with a yellow margin. Leaves are slightly rippled and lightly cupped. Leaves are dull on top and underneath.

Hosta 'Kinakafu Otome'

M. Zilis 2003

[Parentage unknown but we are sure there are genes from *Hosta venusta*]

A small hosta. Leaves are greenish yellow with a medium green margin. Leaves are lightly wavy, dull on top and slightly shiny underneath. The leaves become dark green thru the season

Hosta 'Lakeside Neat Petite'

M. Chastain (R) 1991

[*Hosta venusta* × *Hosta* 'Blue Cadet']

A small blue green hosta. Leaves have large undulations. This hosta can be used as ground cover.

Hosta 'Leather Sheen'

M. Zilis & D. Lohman (R) 1988

[*Hosta* 'Sum and Substance' × *Hosta venusta*]

A medium size fast growing dark green hosta. Leaves are very glossy on top. This hosta looks like *Hosta* 'Marashino Cherry'.

Hosta 'Little Devil'

J. Machen, Jr., 2000

[Sport of *Hosta* 'Swoosh' x *Hosta venusta*]

A small hosta. Leaves are green with a crispy white margin. This hosta can be used in troughs or as ground cover.

Hosta 'Masquerade'

Japon (OI) D. Grenfell (N) AHS, K. Walek (R) 2009

[*Hosta venusta* × sport of *Hosta sieboldii*]

A small hosta. Leaves are white with a green margin. Leaves are slightly rippled and twisted. We must check this hosta carefully because it often returns green with time.

Hosta 'Minuta'

A non registered hosta brought from Japan.

[We are sure there are genes from *Hosta venusta*]

A small fast growing blue green hosta. This hosta looks like *Hosta* 'Thumb Nail' and *Hosta* 'Tiny Tears'.

Hosta 'Paradise Puppet'

Marco Frenzen NR

[Parentage unknown but we are sure there are genes from *Hosta venusta*]

A small non-registered green hosta. Leaves are slightly twisted. This hosta looks like *Hosta* 'Tiny Tears'.

Hosta 'Shining Tot'

P. Aden (R) 1982

[*Hosta venusta* × *Hosta* 'Rock Master']

A small gray green hosta. Leaves are lustrous on top. One of the best miniature hostas. Excellent plant for hybridizing.

Hosta 'Surprised by Joy'

A Malloy (ONIR) 1998

[*Hosta* 'Flamboyant' × *Hosta venusta*]

A small good growing hosta. Leaves are white with a green streaked margin. Leaves are slightly rippled, dull on top and slightly shiny underneath. Petiole is white with a green margin. This hosta can be used in front of a flower bed or in troughs.

Hosta 'Tiny Tears'

R. Savory (R) 1977

[Hybrid of *Hosta venusta*]

A small fast growing medium green hosta. This hosta can be used in front of a flower bed or in troughs.

Hosta 'Tot Tot'

P. Aden (R) 1978

[*Hosta* 'Blue Cadet' × *Hosta venusta*]

A small blue green hosta. This hosta can be used in front of a flower bed or in rock gardens in shady areas.

There are other seedlings in this group:

- *Hosta* 'Alpine Dream'
- *Hosta* 'Amanuma'
- *Hosta* 'Appetizer'
- *Hosta* 'Arrowhead'
- *Hosta* 'Baby Kim'
- *Hosta* 'Ballerina'
- *Hosta* 'Brooklyn's Baby Doll'
- *Hosta* 'Brutus'
- *Hosta* 'Concordia Petite'
- *Hosta* 'Devon Red'
- *Hosta* 'Emoticon'
- *Hosta* 'Evening Gown'
- *Hosta* 'Fairy Ring'
- *Hosta* 'Gemstone'
- *Hosta* 'Golden Cherub'
- *Hosta* 'Golden Gem'
- *Hosta* 'Golden Maid'
- *Hosta* 'Gosan Gold Midget'
- *Hosta* 'Green Fingers'
- *Hosta* 'Harold's Thumb'
- *Hosta* 'Herbie'
- *Hosta* 'Himekomachi'
- *Hosta* 'Honey Moon'
- *Hosta* 'Ichi-gan'
- *Hosta* 'Japangirl'
- *Hosta* 'Kifukurin Otome'
- *Hosta* 'Little Fellow'
- *Hosta* 'Lorna'
- *Hosta* 'Misty Morning'
- *Hosta* 'Nikkou'
- *Hosta* 'Nishikigawa'
- *Hosta* 'Number One'
- *Hosta* 'Ogon Otome'
- *Hosta* 'Orphan Annie'
- *Hosta* 'Otome-no-mai'
- *Hosta* 'Papoose'
- *Hosta* 'Pipin the Short'
- *Hosta* 'Queue'
- *Hosta* 'Red Dwarf'
- *Hosta* 'Red Tubes'
- *Hosta* 'Rock Island Line'
- *Hosta* 'Rock Master'
- *Hosta* 'Shark's Tooth'
- *Hosta* 'Shirofukurin Otome'
- *Hosta* 'Suzuki Thumbnail'
- *Hosta* 'Ted's Friend'
- *Hosta* 'Thumb Nail'
- *Hosta* 'Venucosa'
- *Hosta* 'Venus Star'

To summarize: *Hosta venusta* and the hostas from that group are the smallest of all hostas. They are fast growers. They are not slug resistant. They must be planted in very shady areas. They are very important in hybridizing to get miniature hostas.

Hosta nakaiana species

This species was found during a trip to the Huksan Archipelago in Korea.

This species is very common in shady rocky sites and the sides of North West slopes. In its natural habitat it grows in pine forests among grasses and ferns, in shady and dry regions.

Leaves are elongated, dark green, quite thick, flat and the plant will produce many flowers late in the season.

Plants being sports of *Hosta nakaiana*.

Hosta 'Emerald Tiara'

Walters Gardens, Inc. (R) 1988

[Sport of *Hosta* 'Golden Scepter']

A small hosta. Leaves are yellow with a dark green margin.

Hosta 'Ghost Spirit'

C. Isaacs (ONIR) 1999

[Sport of *Hosta* 'Valentine Lace']

A small hosta. Leaves are green streaked white with a medium blue green margin. Leaves are glaucous on top and underneath.

Hosta 'Golden Scepter'

R. Savory (R) 1983

[Sport of *Hosta* 'Golden Tiara']

A small slown growing hosta yellow gold hosta. This hosta will like the morning sun.

Hosta 'Lemon Frost'

M. Zilis (ONR) M. Zilis & R. Solberg (I) 2002

[Sport of *Hosta* 'Lemon Lime']

A small hosta. Leaves are greenish yellow with a pure white margin. Leaves are heavily rippled, moderately wavy and lightly twisted.

Hosta 'Olive Branch'

S. Moldovan (ONI) P. & J. Ruh (R) 2008

[Parentage unknown but we are sure there are genes from type *naka-ina*]

A small hosta. Leaves are greenish yellow with medium green margin. Leaves are slightly rippled and lightly wavy. Leaves are very shiny on top and underneath. Petiole has a variegated edge.

Hosta 'Twist of Lime'

B. Banyai & R. Solberg (R) 1991

[Sport of *Hosta* 'Lemon Lime']

A small hosta. Leaves are dark yellow with a dark green margin.

Hosta 'Veronica Lake'

F. Riehl (OR) 1993

[Sport of *Hosta* 'Pearl Lake']

A small hosta. Leaves are slightly blue green with a creamy white margin. Leaves are smooth, dull on top and slightly shiny underneath.

There are other sports from *Hosta nakaiana*:

- *Hosta* 'Diamond Tiara'
- *Hosta* 'Duchess'
- *Hosta* 'Emerald Scepter'
- *Hosta* 'Grand Tiara'

As these hostas are extremely fertile, there are many seedlings coming from crossings with *Hosta nakaiana*.

Hosta 'Apple Green'

K. Anderson (R) 1982

[Hybrid of *Hosta nakaiana*]

A small chartreuse hosta.

Hosta 'Blue Boy'

D. Stone & AHS (R) 1986

[*Hosta nakaiana* × *Hosta sieboldiana*]

A medium size blue green hosta.

Hosta 'Blue Cadet'

P. Aden (R) 1974

[Hybrid of *Hosta* 'Tokudama']

A small hosta. Leaves are blue with a white margin. Leaves have slight undulations. We have to be careful with this hosta because many were virussed on the market.

Hosta 'Bountiful'

E. Fisher (R) 1971

[Hybrid of *Hosta nakaiana*]

A small blue green hosta.

Hosta 'Candy Hearts'

E. Fisher (ONI) P. Ruh (R) 1998

[Hybrid of *Hosta nakaiana*]

A small medium blue green hosta. Leaves are slightly rippled, shiny on top and dull underneath.

Hosta 'Drummer Boy'

S. Moldovan (R) 1983

[*Hosta nakaiana* × *Hosta* 'Ruffled Queen']

A small dark green hosta.

Hosta 'Floradora'

P. Aden (R) 1978

[*Hosta nakaiana* × Hybrid of *Hosta longipes*]

A small light green hosta.

Hosta 'Golden Spider'

G. Harshbarger & P. Ruh (R) 1987

[Parentage unknown but we are sure there are genes from *Hosta nakaiana*]

A small bright yellow green hosta.

Hosta 'Golden Tiara'

R. Savory (R) 1977

[Hybrid of *Hosta nakaiana*]

A small hosta. Leaves are green with a very fine white margin.

Hosta 'Lemon Lime'

R. Savory (R) 1977

[Hybrid of *Hosta nakaiana*]

A small chartreuse gold hosta.

Hosta 'Pearl Lake'

Piedmont Gardens (R) 1982

[Parentage unknown but we are sure there are genes from *Hosta nakaiana*]

A small medium green hosta.

Hosta 'Peter Pan'

N. Minks (R) 1980

[*Hosta nakaiana* × *Hosta* 'Helen Field Fisher']

A small grey green hosta.

Hosta 'Valentine Lace'

M. Armstrong (R) 1970

[Hybrid of *Hosta nakaiana*]

A small medium green hosta. Leaves have slight undulations.

Hosta 'Warwick Curtsey'

L.G. Jones (ONI) AHS, K. Walek (R) 2009

[*Hosta* 'Dorothy Benedict' × *Hosta nakaiana*]

A small hosta. Leaves are green with a golden yellow margin. Leaves are lightly corrugated.

There are other seedlings from that group:

- *Hosta* 'Bacchanal'
- *Hosta* 'Betty'
- *Hosta* 'Birchwood Parky's Gold'
- *Hosta* 'Dear Heart'
- *Hosta* 'Dick's Delight'
- *Hosta* 'Dixie Queen'
- *Hosta* 'Goldbrook Genie'
- *Hosta* 'Golden Surprise'
- *Hosta* 'Happy Hearts'
- *Hosta* 'Heartleaf'
- *Hosta* 'Lucky Charm'
- *Hosta* 'Marquis'
- *Hosta* 'Oxheart'
- *Hosta* 'Soft Touch'
- *Hosta* 'Special Gift'
- *Hosta* 'Warwick Edge'

To summarize: hostas from that group were found in their natural habitat where there were pine forests. Normally the plants are small. They must be planted in shady areas. They will tolerate drought periods and they will give many sports.

Hosta yingeri species

This species was also found during a trip in Huksan Archipelago in Korea.

It was growing in the same type of conditions as *Hosta nakaiana*. This species is very common in shady rocky sites and the sides of North West slopes. In its natural habitat it grows in pine forests among grasses and ferns, in shady and dry regions.

Leaves are elongated, dark green and thick. The flowers are pale lavender and appear late in the season. Their thick leaves allow them to retain water like a cactus.

Identified by Makino in 1902

A small erected green hosta. Leaves have heavy substance. This hosta will need moisture and warmth to have a good growth.

This species will not give many sports. The only one known to date is *Hosta 'Gentle Spirit'*.

However there are crossings with *Hosta yingeri* :

Hosta 'Harpoon'

R. Herold (ONIR) 2000

[*Hosta* 'Swoosh' × *Hosta yingeri*]

A small hosta. Leaves are medium green with a creamy white margin. Leaves are slightly rippled and slightly shiny on top and underneath. This hosta looks like *Hosta* 'Carolina Sunshine'.

Hosta 'Korean Snow'

Niche Gardens (O) R. Solberg (NIR) 1999

[Hybrid of *Hosta yingeri*]

A small hosta. Leaves are medium green streaked with creamy white. Leaves are slightly rippled and very shiny on top and underneath. This hosta is quite popular due to the leaves turning green totally along the season.

Hosta 'Lakeside Looking Glass'

M. Chastain (ONIR) 1997

[Hybrid of *Hosta yingeri*]

A small dark green hosta. Leaves are slightly rippled and very shiny on top and underneath. Petiole has blue black sheen.

Hosta 'Potomac Pride'

T. Avent (OR) 1995

[*H. yingeri* × *H.* 'Blue Umbrellas']

A large medium blue green hosta. Leaves are smooth, very shiny on top and glaucous underneath. The first time this hosta was seen during an American Hosta convention in 1995 as it was given to all attendants.

Hosta 'Quill'

R. Herold (ONI) AHS, K. Walek (R) 2009

[*Hosta pycnophylla* × *Hosta yingeri*]

A small medium blue green hosta. Leaves are wavy and twisted. This hosta looks like *Hosta laevigata*.

Hosta 'Swamp Thing'

T. Avent (ONIR) 2005

[A blue seedling from *Hosta sieboldiana* x a seedling of *Hosta yingeri*]

A small hosta. Leaves are medium green with a golden yellow margin. Leaves are slightly rippled, lightly folded and moderately corrugated. Petiole has variegated wings.

There are other seedlings in that group:

- *Hosta* 'Blue Haired Lady'
- *Hosta* 'Celtic Uplands'
- *Hosta* 'City Slicker'
- *Hosta* 'Crystal Chimes'
- *Hosta* 'Green Jeans'
- *Hosta* 'Jaz'
- *Hosta* 'Kermit'
- *Hosta* 'Kiwi Fresh Salad'
- *Hosta* 'Kiwi Leap Frog'
- *Hosta* 'Mystic Star'
- *Hosta* 'Old Coot'
- *Hosta* 'Runaround Sue'
- *Hosta* 'Southern Gold'
- *Hosta* 'Sun catcher'
- *Hosta* 'Sunny Side Up'
- *Hosta* 'Sweet Tater Pie'
- *Hosta* 'Terpsichore'
- *Hosta* 'Whiskey Sour

Hosta laevigata species

This species is totally like *Hosta yingeri*, except for the leaves which are longer, narrower and thicker. This hosta doesn't produce many sports. Seeds are fertile but take more time to produce.

Hosta laevigata

This species gives small plants with elongated shiny green leaves. This hosta is a bit bigger than *Hosta yingerii*. If used for hydridizing, this hosta will pass on the corrugated leaves.

Hosta 'Ray of Hope'

A Tower (O) A. Tower & R. Solberg (NI) R. Solberg (R) 2005

[Sport of *Hosta laevigata*]

A small hosta. Leaves are medium green with streaks of cream throughout. Leaves are heavily rippled and intensely wavy. Petiole is wide and wavy. This hosta will turn more yellow during the season.

Hosta 'Roller Coaster Ride'

M. Zilis (OR) R. Solberg (N) M. Zilis & R. Solberg (I) 2005

[Sport of *Hosta* 'Ray of Hope']

A small hosta. Leaves are medium green with a creamy white margin. Leaves are heavily rippled and slighty shiny.

To summarize: hostas from this group were found in their natural habitat where there were pine forests. Normally the plants are small. They must be planted in shady areas. They will tolerate drought periods because of their thick leaves like a cactus.

Hosta capitata species

This species naturalized in Japan is almost the same as *Hosta nakaiana*, from Korea, except for the leaves are more elongated, narrower and thicker. This species doesn't produce many sports. Seeds are fertile but slow to produce.

Identified by Makino in 1902. A small erected green hosta.

The only sport known so far is *Hosta* 'Gentle Spirit'.

Hosta capitata

If we summarize: Hosta capitata is droughts tolerant. It should be planted in shaded areas. These hostas will bloom in sunny areas but need more watering. If planted in half shaded areas (hald day of sun) add peat moss.

There are a few more species that could be studied for their culture habit like *clausa, gracillina, longipes, minor, rectifolia and tardiva* but these are very rare and reserved for specialist.

There are many more hostas that are not part of all the species described herein.

Chapter 5

The most popular hostas

Each year the American Hosta Society conducts a popularity poll among its members to determine their favorite hostas. It is very surprising to see, years after years, the hostas appearing on the list. Hosta growers selected those hostas as they like their beauty and their facility to grow. This is another good way to make sure succeeding with their culture.

Marco Franzen from Holland is one of the best hostas growers in Europe.

Hosta 'Sagae'

This hosta was registered in 1996 by l'AHS. It is a large vase-shaped hosta with leaves being yellow green with bright pale yellow margin. This hosta was the Hosta of the Year in 2000. This hosta is slug resistant.

Hosta 'June' almost always on top of the list.

This sport of *Hosta* 'Halcyon' was registered by Neo Plants in 1991. This hosta has leaves that are light yellow with a dark blue margin. This hosta can be planted full sun or in shady areas. However this hosta may have looks very different depending where planted. This hosta is sun burn and slug resistant.

Hosta 'Liberty'

This large hosta was registered in 2000 by John Machen. Leaves are medium blue green with a golden yellow margin. This hosta is a sport of *Hosta* 'Sagae'. Leaves are thicker making it slug resistant.

Hosta 'Sum and Substance'

A very large hosta registered in 1980 by Paul Aden. It is probably the largest hosta I have seen so far. Leaves are waxy chartreuse gold. Leaves are thick so it is slug resistant. This hosta should have a good spot in a garden.

Hosta 'Niagara Falls'

This large hosta was registered by Bill Brinka and Olga Petryszyn in 1991. This hosta has thick dark green leaves that have slight undulations. This hosta is considered as one of the best hosta with green leaves. Another slug resistant hosta.

Hosta 'Guacamole'

This large hosta was registered by Bob Solberg in 1994. The thick leaves are golden yellow with a dark green margin. Being in the *Hosta plantaginea* line, the flowers are almost white and fragrant. This hosta is sun burns resistant.

Hosta 'Guardian Angel'

This large hosta was registered by C. and R. Thompson in 1995. The thick leaves are light green with an intensely blue green margin. With the corrugated and twisted leaves, this hosta is slug resistant.

Hosta 'Paradigm'

This large hosta was registered in 1999 by Walden-West. The leaves are light green with a dark green margin. Leaves are thick and corrugated, so it is slug and sun burns resistant. This hosta was named Hosta of the Year in 2007.

Hosta 'montana Aureomarginata'

This large hosta was registered by AHS in 1987. Leaves are dark green with a yellow margin. Leaves are undulated.

Hosta 'Blue Angel'

Another large hosta registered by Paul Aden in 1986. Very large blue leaves that are slug resistant. It can tolerate few hours of the morning sun.

Hosta 'Sun Power'

A large erected form hosta registered by Paul Aden in 1986. Leaves are light chartreuse with large undulations. This hosta is considered as the the best yellow hosta. This hosta is sun burns resistant.

Hosta 'Krossa Regal'

This large size hosta was registered by G. Krossa and A. Krossa (R) in 1980. This hybrid of *Hosta nigrescens* has leaves that are blue gray with large undulations.

Hosta 'First Frost'

This medium size hosta was registered by Scolnik and R. Solberg in 2002. Sports of *Hosta* 'Halcyon', the leaves are blue green with a yellow margin. The margin turns white along the season. Leaves always stay nice.

Hosta 'Cathedral Windows'

A large hosta registered by Hans Hansen in 2005. Sport of *Hosta* 'Stained Glass', leaves are yellow with a green margin. This hosta is very nice in spring and nicer in summer.

Hosta 'Paul's Glory'

A large hosta registered by Peter Ruh and Hofer in 1987. Leaves are yellow with irregular blue green margin. Even if this hosta is sun tolerant, the center of the leaves will be pale yellow in full sun. The thick leaves make it slug resistant.

Hosta 'Striptease'

A large hosta registered by C. and R. Thompson in 1991. This hosta has leaves that are very special: there is a leaf pattern in the middle of the leaves. Leaves are chartreuse yellow with dark green margin. There is fine white line around the leaf design. This hosta is resistant to sun burns.

Hosta 'Praying Hands'

This medium size erected hosta was registered by G. Williams in 1996. Leaves are dark green with a very fine yellow gold margin. Foliage is very unusual being heavily rippled. This hosta is also slug and sun burns resistant.

Hosta 'Abiqua Drinking Gourd'

This large blue hosta was registered by Walden West in 1987. Leaves are dark blue, corrugated and in cupped shape. After a rain fall or watering, leaves will keep water that attracts birds. A slug resistant hosta.

Hosta 'Atlantis'

This ¸very large hosta was registered by Tony Avent and Hans Hansen in 2004. Leaves are dark green with a golden yellow margin. Leaves are slightly rippled and lightly wavy.

Hosta 'Great Expectations'

This large hosta was registered by Paul Aden in 1988. Leaves are chartreuse yellow with a dark green margin. This hosta needs the morning sun followed by little dappled shade to perform at its best. This hosta is slug resistant. A good protection with mulch in autumn will allow winterizing the growth for spring.

Hosta 'El Niño'

This medium size hosta was registered by the Hollander Warmerdam in 2003. Leaves are blue green with a pure white margin. This hosta is a sport of *Hosta* 'Halcyon'. It is sun burns resistant.

Hosta 'Regal Splendor'

This large erected hosta was registered by Walters Gardens in 1987. Leaves are bluish gray with a creamy white to creamy yellow margin. This hosta is slug resistant.

Hosta 'Spilt Milk'

This large hosta was registered by Mildred Seaver in 1999. The thick leaves are slightly blue green streaked with white. A slug resistant hosta. As the name says this hosta look like it was **splashed with milk.**

Hosta 'Brother Stephan'

This medium size erected hosta was registered by Olga Petryszyn in 1998. Leaves are yellow with a medium green margin. With its slightly rippled leaves this hosta reaches this list for the first year.

Hosta 'Blue Mouse Ears'

This miniature hosta was registered by Deckert in 2000. This is the only miniature hosta reaching this top list. As you can see all the other ones are quite large hostas. As soon this hosta came on the market, it was a must for every hosta collector. We will talk about this in a chapter on miniature hostas.

If you collect these hostas, you can be assured to succeed. Even more almost nothing to do after planting: most of them are sun and slug resistant.

Chapter 6

Miniature hostas

Space in your garden is very limited to plant new hostas. There are numerous miniature hostas. Even more, if you have in your garden, a large hosta that you don't like, just remove it and you will have room to plant many small hostas.

Miniature hostas will be between 4 to 8" (10 to 20 cm) in diameter.

You can plant some between your other hostas, but we are not in favor of that practice. Try to plant them together in the same flower bed.

Today we can find miniature hosta in almost all colors, the same as the larger ones. We have access to few hundreds of miniature hostas.

Your hostas will be all at the same scale sizes. You can also plant slow growing conifers and few small companion plants.

If you want to be certain that visitors don't step on your miniature hostas, plant them in a raised bed. Even more the visitors won't have to bend to look at your famous specimens.

A good method is to use rocks, walls or special structures to create your miniature hosta beds.

These structures were specially made for miniature hostas.

Plant tags are always a problem. Be careful! Many times plant tags are more visible than the plants themselves.

You can also plant your miniature hostas in containers. There are troughs but many other types.

During annual conventions with the American Hosta Society, members present containers arrangements.

Planting and soil

Miniature hostas need more care when planting. Soil must be loose and have good drainage. Planting soil, peat moss and coarse grit are best for drainage. When planting choose a lighted site and leave at least 6" (15 cm) between each hosta.

In your miniature hosta bed, add slow growing conifers and alpine plants.

If you plant miniature hostas in containers, make sure soil is not compacted. Containers must not be allowed to dry out. They must be watered more often.

Miniature hostas have to be protected where rough winters. Frosts and thaws might kill them. If they are already in the ground add mulch (compost, shredded leaves, peat moss, etc.). The plants in containers would be better buried in the ground for a good winter survival.

Shredded leaves mulching on miniature hostas

Popularity

Miniature hostas are very popular in landscapings.

Every year, the American Hosta Society runs a popularity poll among its membership to find out which miniature hostas are the most popular ones. Many hundreds of them can be found on the market today.

1. *Hosta* 'Blue Mouse Ears'

2. *Hosta* 'Pandora's Box'

3. *Hosta* 'Cameo'

4. *Hosta* 'Dragon Tails'

5. *Hosta* 'Frosted Mouse Ears'

6. *Hosta venusta*

7. *Hosta* 'Tiny Tears'

8. *Hosta* 'Cherish'

9. *Hosta* 'Cat and Mouse'

10. *Hosta* 'Popo'

There are hundreds of these miniature hostas now available to gardeners.

Hosta 'Cat's Eye'

Hosta 'Cody'

Hosta 'Awesome'

Hosta 'Cookie Crumbs'

Hosta 'Bitsy Gold'

Hosta 'Country Mouse'

Hosta 'Cheatin Heart'

Hosta 'Dixie Chick'

Hosta 'Faithful Heart'

Hosta 'Hope'

Hosta 'Green Eyes'

Hosta 'Hush Puppie'

Hosta 'Gold Drop'

Hosta 'Hydon Sunset'

Hosta 'Green Mouse Ears'

Hosta 'Itsy Bitsy Spider'

Hosta 'Hart's Tongue'

Hosta 'Lakeside Down Sized'

Hosta 'Lemon Lime'

Hosta 'Masquerade'

Hosta 'Little Aurora'

Hosta 'Medusa'

Hosta 'Little Caesar'

Hosta 'Royal Mouse Ears'

Hosta 'Little Razor'

Hosta 'Saishu Jima'

Hosta 'Little Wonder'

Hosta 'Sitting Pretty'

Hosta 'Stiletto'

Hosta 'Surprised by Joy'

Hosta 'Teeny Weeny Bikini'

Hosta 'Tot Tot'

Hosta 'Yellow Waves'

Hosta 'Twist Of Lime'

Hosta 'White Swan'

Hosta 'X-Rated'

Hosta 'Yellow Eyes'

Hosta 'Illicit Affair'

As we can see, growing hostas is more accessible these days.

Chapter 7

Landscaping with hostas

As we know now, there is a large variety of hostas, we also know where to plant them to obtain best results, let's see how to use them. Of course there are multiple opportunities, we could write another book just on landscaping and companion plants.

I will only comment few landscapings that I have seen while visiting gardens.

We utilize the cavities in that tree to place hostas in containers.

Hostas are always at their best along a pond, a stream or a cascade.

In a flower bed, you can plant hostas in all sizes and colors.

Always plant an erected shape hosta in the background.

Some hostas in containers were set up among different hostas in a landscape.

Ornamental grasses and perennials will light up any hosta beds.

All around a tree, hostas are great plants to use.

A long and wide hosta bed along a pathway.

Only one hosta in the middle of conifers, perennials and annual plants.

This hosta lover has over 500 different hostas in this garden in Ontario.

You can plant all types of hostas together: different in colors and sizes.

Miniature hostas are planted in a raise bed.

Gardening and Hosta Organisms

Many organizations are existing to promote plants and hostas. If you want to know more and want to join a Gardening Club near your home, in Province of Quebec, Canada, you can consult:

Quebec Garden Clubs Federation (F.S.H.E.Q.)

4545, Pierre-De Coubertin avenue
P.O. Box 1000, branch M
Montreal, (Quebec)
H1V 3R2
Phone: (514) 252-3010
Email: fsheq@fsheq.com
Website : http://www.fsheq.com/

Other organisms are specialized only with hostas. All activities from these Societies are about these plants and their companion plants. These Societies can supply you with all information on the culture, exchange and visits of display gardens, public or private.

In Quebec there is Quebec Hosta and Hemerocallis Society, has many hundreds of members. You can reach them at:

Quebec Hosta and Hemerocallis Society (Q.H.HS.)*
4101, east Sherbrooke street
Montreal, (Quebec)
H1X 2B2
Phone : 514 892-5388
Email : sqhh-qhhs@live.ca
Website : https://sites.google.com/site/hostaquebec/

Q.H.H.S. is a member of the American Hosta Society, Society with thousands of hosta lovers, not only in United States but all over the world. They have also the mandate to keep in order the registration of the gender *Hosta*. The organism publishes many documents on hostas. You can reach them at

American Hosta Society (A.H.S.)
C/O Sandie Markland
AHS Membership Secretary
Post Office Box 7539
Kill Devil Hills, NC 27948
Website: http://www.americanhostasociety.org/

The American hosta growers have also their Association:

The American Hosta Growers Association
c/o Bob Solberg,
Executive Secretary AHGA
P.O. Box 16306
Chapel Hill, NC 27516
Phone: 919-309-0649
Email: greenhill@mindspring.com
You can also consult their website: http://www.hostagrowers.org/

Bibliography and Suggested Readings

English

- *The Genus Hosta – List of Registered Cultivars, 1st Edition – May 1993 (American Hosta Society)*
- *The Genus Hosta – List of Registered Cultivars, Supplement I – January 2000 (American Hosta Society)*
- *Registration 2001 – The Hosta Journal Volume 33 Number R (American Hosta Society)*
- *Garden Judging and Award Guide (American Hosta Society) with a revised list by W. George Schmid*
- *The American Hosta Guide – The 2002 catalog from Wade & Gatton Nurseries*
- *The Genus Hosta W.G. Schmid, Published in 1992 by Timber Press Inc*
- *The Hosta Book P. Aden, Published in 1988 by Timber Press Inc*
- *Hosta : The Flowering Foliage Plant D. Grenfell, Published in 1990 by Timber Press Inc*
- *The Gardener's Guide to Growing Hostas D. Grenfell, Published in 1996 by Timber Press Inc*
- *The Hosta Handbook M. Zilis, Published in 2000 by Q & Z Nursery Inc*
- *Hostas and other shade-loving plants, R. Bird and D. Tarrant Published in 1999 by Whitecap Books Ltd*
- *The Book of Little Hostas Michael Shadrack, Kathy Guest Shadrack, published in 2011, Timber Press*
- *The Hostapedia: a hosta encyclopedia Mark Zilis, published in 2009, Q & Z Nursery*
- *Hostas, an essential guide Richard Ford, published in 2010, The Crowood Press*
- *The new encyclopedia of Hostas Diana Grenfell and Michael Shadrack, published in 2010, Timber Press*

- *Pocket Guide to Hostas Diana Grenfell and Michael Shadrack published in 2007 Timber Press*
- *Hostas Rosemary Barrett, published in 2004, David Bateman Ltd*
- *Hostas Andrew Mikolajski published in 2001, Southwater Publishing*
- *The Hosta Handbook - Second Edition, M. Zilis published in 2013, Q & Z Nusery inc.*

French

- *Les Hostas Réjean D. Millette, published in 2003 by Les Éditions de l'Homme*
- *Les Hostas Jane Phillips, published in 2006, Le Jardin anglais*
- *Les Hostas Réjean D. Millette, published in 2013, by Les Éditions du Sommet*

German

- *Hosta* Fritz Köhlein, published in 1993, Ulmer Eugen Verlag

- *Hosta Funkie* Sandra Bond, published in 1993, Haymarket Media Gmbh & Co

Index